LIFE UNIVERSITY
A DAILY DEVOTIONAL BOOK FOR
COLLEGE STUDENTS & YOUNG ADULTS

BY

JAY STEWART & JOHNNY JERNIGAN

SOAR PRODUCTIONS

This book is dedicated to the preservation of America's greatest resource, its young people, in the hope that they will stay strong in the LORD through these tumultuous years.

Cover design by Doug Johnson

All Scripture quotations are taken from the New International Version of the Bible.

LIFE UNIVERSITY
A Daily Devotional Book For College Students and Young Adults

Copyright © 1994 by SOAR Productions
1-800-251-1336

Published by SOAR Productions

JANUARY 1   ON YOUR MARK, GET SET, GO!

Welcome. Welcome to a new day, a new year, new beginnings. Today marks the day that millions of people will evaluate life and self and determine that "things are gonna be different." New Year's resolutions come in all shapes and sizes, much like the people who make them. Maybe your own self-inventory and the work of the Holy Spirit revealed some areas in your life that need work. Poor grades last semester, being out of shape, a drinking problem, bitterness, lack of discipline in your Christian walk, or sexual misconduct are only a few reasons people seek change.

The solid New Year's counsel Hebrews gives really hits home. It says in Hebrews 12:1, *"... let us throw off everything that hinders and the sin that so easily entangles, and let us run with perseverance the race marked out for us."* That really sums up all New Year's resolutions. Let's get down to business, God's business, and determine to be champions for Him. God will help us, and the result will be the best year ever!

JANUARY 2   THE POWER OF PERSUASION

Read Judges 16:15-22. There are many situations each day that can persuade us to do the right thing or the wrong thing.

According to an executive of an Oakland California company, all employees are required to sign in when they come to work each morning, and, if late, to write down the reason. During bad weather, the reason is generally "fog" — written in by the first arrival, with those who come in afterwards simply putting ditto marks. One morning the first tardy gentleman printed under the excuse column, "Wife had twins." On the next 18 lines, the later arrivals dutifully noted ditto.

Are those around you today encouraging and persuading you to live for Christ or to do the wrong things? Write down how you can persuade others today to live for Jesus!

In a recently published book entitled "Laugh Again," Charles Swindoll urges people to find outrageous joy in the midst of difficult days. Someone once said that if you don't learn to laugh at trouble, you won't have anything to laugh at when you're old. Life is certainly not easy nor pain free. No college or university campus is exempt from turmoil or hardships that can greatly affect our level of joy each day.

The University of Michigan research discovered that the average child laughs 150 times a day; the average adult, 15 times. Somehow, between childhood and adulthood, we allow our joy to be stolen from us. But joy is a choice of our will, not an emotion dependent on circumstances. In the midst of adverse circumstances, Paul writes in Romans 15:13, *"May the God of hope fill you with all joy and peace as you trust in him..."* Paul didn't pray God would supply more money, or a better tent, or more cattle. He asked God to give the believers joy. Pray and ask God today for His joy, which is your strength.

At an educational conference for secondary education teachers and administrators, a college professor was approached by a man who asked, "Do you remember me?"

"Sure," replied the professor unconvincingly.

"What did I do in college?" he challenged.

The professor replied, "You did absolutely nothing."

Surprised, the man said, "You do remember me!"

What do you want to be known for? In Revelation 2:2, God pays a high compliment to the church in Ephesus. He says, *"I know your deeds, your hard work and your perseverance."* Does God know you because of your hard work? Dedication and hard work while in school may earn us a degree and a gold tassel. Hard work for a corporation may bring a promotion or a plaque. These things will eventually fade away. And all that will remain is what we have accomplished for God's kingdom. Today is a new chance to do that something that really matters.

JANUARY 5   *IN GOOD HANDS*

Isaiah 52:12 says, *"The Lord will go before you, the God of Israel will be your rear guard."*

When you're a senior in high school, you're supposed to feel invincible. Everybody says you have your whole life ahead of you and can be anything you want to be—even the president. Right? Then how come a simple little question from your aunt like, "By the way, what are you going to do next year?" causes a major panic attack?

Far from feeling invincible, you feel as if you're about to be swallowed by the great big world you're supposed to have by the tail. What are you going to do with your life? Marriage? College? Career? You haven't the foggiest, and the fact alone scares you to death.

While you can't know where the future will lead, God can because He has scouted up ahead. *"The Lord will go before you,"* Isaiah says in the verse above. You need not worry. *"For I know the plans I have for you,"* declares the Lord, *"plans to prosper you and not to harm you, plans to give you hope and a future"* (Jeremiah 29:11).

If that's not security enough, Isaiah also offers the assurance that God follows behind you: *"The God of Israel will be your rear guard."* This is military lingo. The Israelite army always had a vanguard and a rear guard. The vanguard preceded the soldiers and scouted unexplored territory. The rear guard followed behind, helping stragglers, and more or less picking up the pieces. God does the same thing. He understands that you are sometimes hesitant to move forward because of past failures, bad memories, nagging sins, and old wounds. No problem. God follows behind, forgiving your sins and helping you back on your feet. He picks up the pieces of your life, mending and restoring hurts and sorrows, and even creating positive results from negative incidents.

Indeed God has a magnificent plan for your life that exceeds your wildest imaginings. And He's willing to both guide and guard you toward that destination. Why? He simply wants you to experience the "riches of his glory," which He has planned for you from the beginning of time.

Ask God how you can walk withHim today.

JANUARY 6   WILL YOU ANSWER THE CALL?

"Call to me and I will answer you and tell you great and unsearchable things you do not know" (Jeremiah 33:3).

Telephones are a necessary part of everyday life; so are voice pagers, cellular phones, desktop computers and computer phone calls.

Several years ago, this ad appeared in an English publication for six men to go to China:

> Six courageous young men are needed at once to go to China for the Chinese Industrial Co-Ops which trains technicians for a democratically industrialized China. They are to take the place of George Hogg, brilliant 31-year old Oxford graduate, who died of tetanus in Northwest China beyond reach of medical aid. If willing to risk disease, endure discomfort, eat only Chinese food, talk Chinese, he may apply immediately at the Anglo-Chinese Development Society. Anyone not prepared to take similar risks need not apply.

Over 600 young men applied! God is calling out today for men and women to stand for righteousness in our colleges and universities. Write down how you can answer the call.

Over and over, throughout God's word we see that God's plan has been to search for man. God doesn't track man down through AT&T, but in Revelations we find that God makes house calls. In Revelation 3:20, we read, *"Here I am! I stand at the door and knock. If anyone hears my voice and opens the door, I will come in..."* WOW! Maybe you know what it is like for God to knock on your heart's door, and then to invite Him into your life.

Unfortunately, some people's lives resemble the story of Amber, a young teen. She was so excited to finally get her own phone in her room. But one day, her father walked in to find Amber sobbing in the middle of a sea of debris. "My phone just rang," she cried. "I heard it, but I couldn't find it." Don't allow your life to become so busy and cluttered that you miss God's house call as He knocks at the door of your heart.

JANUARY 8   THE PRODUCT OF PERMISSIVENESS

"So Abram said to Lot, 'Let's not have any quarreling between you and me, or between your herdsmen and mine, for we are brothers. Is not the whole land before you? Let's part company. If you go to the left, I'll go to the right; if you go to the right, I'll go to the left" (Genesis 13:8-9).

In America, all things are permissible. The inscription on the Plymouth Rock monument is a challenge to every generation of Americans: "This spot marks the final resting place of the Pilgrims of the Mayflower. In weariness and hunger and cold, fighting the wilderness and burying their dead in common graves that the Indians should not know how many had perished, they laid the foundations of a state in which all men for countless ages should have liberty to worship God in their own way. All you who pass by and see this stone, remember, and dedicate yourselves anew to the resolution that you will not rest until this lofty ideal shall have been realized throughout the earth."

The Mayflower was permitted to look for a new land—and found it. In college life, all is permissible—but not all is good. What are you permitting yourself to see, hear, and do today? Are you finding freedom or bondage? Lot found destruction. Write down how, through permission, you can find freedom today.

JANUARY 9 — NEED A LIFT?

If there is one product or name that is recognized around the world, it would have to be Coca Cola. When Coca Cola was first invented back in 1886 in Columbus, Georgia, it was marketed by its inventor, Dr. John Pemberton, as a "pick–me–up". It was an elixir or sorts. Over one hundred years later, people world–wide are still in search of a pick–me–up that will carry them through rough times and through each day. Drugs such as crack cocaine, new forms of entertainment, and new breakthroughs in health and medicine have been introduced and tried by many.

James describes a 100% effective, tried and true pick–me–up in James 4:10. He says, *"Humble yourselves before the Lord, and he will lift you up."* Sound too simple? It is. All that is required is that we admit to God we need His help. And He will pick us up spiritually out of sin, physically out of sickness, or emotionally out of depression. All of the world's solutions are temporary. When Jesus lifts you up, it is permanent!

JANUARY 10 — STEP INTO THE WATER

Do you sometimes feel as though you are drowning? Read Genesis 6:11-18.

I am reminded of the great storm of Sidmouth, and of the conduct of the excellent Mrs. Partington on that occasion.

In the winter of 1824 there set in a great flood upon that town—

the tide rose to an incredible height–the waves rushed in upon the houses–and everything was threatened with destruction. In the midst of this sublime storm, Dame Partington, who lived upon the beach, was seen at the door of her house with mop and patterns, trundling her mop, and squeezing out the seawater, and vigorously pushing away the Atlantic Ocean. The Atlantic was roused; Mrs. Partington's spirit was up; but I need not tell you that the contest was unequal. The Atlantic Ocean beat Mrs. Partington. She was excellent at a slop or a puddle, but she should not have meddled with the ocean tempest.

There is such a flood of relativism on most campuses today. It seems we only make a dent or swipe with a mop! God delivered Noah through the flood and He will do the same for you. Write down how you can weather the storm today.

JANUARY 11   GLORIOUS GRADUATION

What kind of thoughts run through your brain when you hear the word "graduation?" Whether you recently graduated from high school or you are now anticipating graduation from college, it is an event like none other. You live for the day to walk that aisle, grab the sheepskin, and move the tassel. You work, slave, sweat, pray, and beg your way through to the final goal of graduating, knowing that life will really begin after that. But so often, after the echo of "Pomp and Circumstance" fades and the cheers, gifts, and parties are gone, all that is left is loneliness, fear and disappointment.

The Bible speaks of a graduation that will far exceed those we have experienced on earth. And there will be no disappointment, fear, or loneliness. In I Thessalonians 4:17 it says, *"...we who are still alive and are left will be caught up together with them in the clouds to meet the Lord in the air. And so we will be with the Lord forever."* We will graduate from this life to the ultimate party, the Kingdom of Heaven!

Read Exodus 7:1-5. It's all around us today–billboards, TV, talk shows, Hollywood and red ribbons.

The *Saturday Evening Post* gave this report of the influenza epidemic at the close of WWI:

"No recorded pestilence before or since has equalled the 1918-1919 death toll in total numbers. In those years, an estimated 21,000,000 died of influenza-pneumonia throughout the world, some 850,000 in the United States alone." More lives were lost in the flu epidemic than on the battlefields of World War I.

Many experts predict that more people will die as a result of the AIDS virus than all deaths in wars from 1901-1991. Death is all around us. The Bible says, *"It is appointed unto man once to die and after that to face the judgement."* The deliverance from aids is not a vaccine but a voluntary surrender of our life to Christ! Are you ready to die? It's more than abstinence. It's an absolute resolve in your relationship with Jesus Christ! Write down how you can live in purity for Christ today.

In Colossians 2:8, Paul issues a warning that all Christians should be careful to heed, especially in this day and age. He says, *"See to it that no one takes you captive through hollow and deceptive philosophy..."* It is imperative that we stay on our toes and stay grounded in God's Word. Many people today present ideas that are partial truths from God's Word mixed with their own opinions or philosophies. Alas, a cult is born.

Some deceptive devices are not as obvious as the one reported by a Scripps Howard News Service reporter. It seems that a Rev. Ewing, pastor of the "Church by Mail", instructed his congregation to send him $18.19--for Matthew 18:19--and in return he would send two plastic, adhesive-backed Bibles. The people were to stick the plastic Bible to an ailing body part and pray for God's healing. Or they could stick theBible to their billfold and watch the blessings pour in. Talk about hollow and deceptive devices! Don't be deceived.

JANUARY 14   CLASS STARTS AT 9:00AM SHARP

Getting to class on time is sometimes a real adventure...traffic jams, oversleeping. Read Exodus 12:31-36.

Some years ago, a young woman by the name of Wilson, who lived near Philadelphia, was capitally convicted of crime, and condemned to die. The day of execution was appointed. In the meantime, her brother used his utmost efforts to obtain a pardon from the governor. He at length succeeded, and hastened to save his sister.

His horse foamed and bled as he spurred it on; and there was no doubt of his succeeding, but an unpropitious rain had swelled the stream: he was compelled to pace the bank, while his heart was ready to break, as he gazed upon the rushing waters that threatened to blast his only hope. The very moment that a ford was at all practicable, he dashed through the river, and arrived at the place of execution. But alas! He was too late, and could only witness the last struggles of his sister on the fatal scaffold.

We must be on time to meet Jesus! Now is the time to go as the Israelites and set out for your promised land. Write down how you can be on time for Jesus today.

JANUARY 15   THE MAIN ENTRANCE

Three men had adjacent businesses that were located in the same building. The businessman on one end of the building put a sign over his store that read, "Year–End Clearance Sale." At the other end, the businessman who ran that store followed with a sign that read, "Closing–Out Sale." The man in the middle knew his business would suffer greatly, so he put up a sign that read, "Main Entrance."

Many different religions offer some pretty attractive bargains. However, fulfillment is found not in religion, but in a relationship with Jesus. He is the Main Entrance to abundant life and eternal life. Jesus said of Himself in John 10:9, *"I am the gate; whoever enters through me will be saved. He will come in and go out, and find pasture."*

JANUARY 16   FATHER KNOWS BEST

Fathers come in all shapes and sizes–tall, short, thin, heavy, bald. Today, 6 out of 10 marriages end in divorce. "Father" has taken on a whole knew meaning.

It happened some years ago that a most urgent and unusual invitation came to me to visit a military academy, in which the students had mutinied, in the hope that possibly I might be of service to the situation. The students had struck in everything: lessons, study hours, drill. The principal handed me a large number of telegrams which had come from the parents who had been wired regarding the situation. These messages were telescopes through which one could look into the various kinds of boys' homes and the parental relationships connected with them.

One father wired his son, "I expect you to obey." Another said, "If you are expelled from school, you needn't come home." Still another, "I'll send you to an insane asylum if you are sent home." Another said, "I'll cut you off without a shilling if you disgrace the family." But the best message was couched in these laconic words: "Steady, my boy, steady! Father."

Even though your father may disappoint you from time to time, your heavenly Father will never let you down. Remember, Father knows best. Write down how you can love your earthly and heavenly Fathers today.

The word apathy is defined as "absence of emotion." A person who is apathetic shows neither love nor hate, neither sorrow nor joy, neither like nor dislike. A man courted his dream girl for months, lavished her with gifts, and went above and beyond the norm in expressing his love to her. On a special moonlit night, he asked her this question: "Do you love me as much as I love you?" She answered, "I don't love you, but I don't hate you either." Yuk!

That is worse than taking a drink of lukewarm water on a hot summer day. Maybe that is why God said in Revelations 3:16, *"So, because you are lukewarm–neither hot nor cold–I am about to spit you out of my mouth."* When we are apathetic towards a God who has expressed His love for us by lavishing us with His goodness, it makes God want to vomit. Let's get hot for God! How do we do it? By learning who He is and daily committing our lives to Him.

JANUARY 18   STRENGTH IN NUMBERS

Some people feel alone in a crowd. Often a crowd breeds security. Read I Samuel 14:1-15. Strength can be found in numbers.

In 1924, when Notre Dame went to Princeton, Knute Rockne had such a sore throat that he could not deliver his pre-game oration. In the adjoining room, the Irish could hear Roper exhorting Princeton to fight valiantly and well.

"There's the best pep talker in the world," Rockne told his men, nodding to the next room. "Listen to him and win with his fight talk." They did.

The sound of the crowd can draw you closer to Christ or push you farther away! Are you lonely in the crowd today? Ask yourself, "Are my friends helping me or hurting me?" Write down your answer today.

Do you want to be set free? We're talking totally free from past guilt and sins. Read Romans 8:1. So many people today want to live for God and move on in life, but the guilt from past sins creates a heaviness, kind of like trying to swim while wearing a dress suit and dress shoes. This scripture says, *"Therefore, there is now no condemnation to those who are in Christ Jesus."*

Some reruns are fun to watch on TV, like those of Barney Fife and Andy, Opie and Aunt Bea, or *Ralph and the Honeymooners*. But others are sheer torture, like *Room 222* and *The Mod Squad*. Some of the memories we have in life are great. But when the devil begins trenching up some of the mistakes we have made, it's a drag. And we don't have to watch. He only wants to hang a "condemned" sign around our neck and make us think we are good for nothing. But God says there is no condemnation for the Christian. Those old sins have been put under the blood of Jesus. And remember, the next time Satan reminds you of your past, you remind him of his future!

Home Alone was a blockbuster hit movie in the early 1990's. Many students feel they are all alone when they are away at school. Read II Kings 6:15-17.

From childhood Albrecht Durer wanted to paint. Finally, he left home to study with a great artist. He met a friend who also had this same desire and the two became roommates. Both being poor, they found it difficult to make a living and study at the same time. Albrecht's friend offered to work while Albrecht studied. Then when the paintings began to sell, he would have his chance. After much persuasion, Albrecht agreed and worked faithfully while his friend toiled long hours to make a living.

The day came when Albrecht sold a wood-carving and his friend went back to his paints, only to find that the hard work had stiffened and twisted his fingers and he could no longer paint with skill. When Albrecht learned what had happened to his friend he was filled with great sorrow. One day returning home unexpectedly he heard the voice of his friend and saw the gnarled, toilworn hands folded in prayer.

"I can show the world my appreciation by painting his hands as I see them now, folded in prayer." It was this thought that inspired Albrecht Durer when he realized that he could never give back to his friend the skill which had left his hands.

Durer's gratitude was captured in his inspired painting that has become world famous. And, we are blessed by both the beauty of the painting and the beautiful story of gratitude and brotherhood.

Your friends in Christ can help you see the victory. Call or write a friend today and let him know how he has helped you.

JANUARY 21   IT ONLY TAKES A SPARK

It all began with a spark from a discarded cigarette, a campfire, or a lightning strike. The end result was total destruction. Millions of acres of forest land were burned in southern California. The fire burned out of control for weeks, gobbling up trees, million dollar homes, animals, and even human lives. What devastation!

In James 3:5, we are asked to ponder or consider how a small spark can so such damage. James was not a member of the Smokey the Bear Club, nor was he the "Father of Forestry." He is comparing the damage a small spark can do to the damage the human tongue can do. He says, *"Likewise the tongue is a small part of the body, but it makes great boasts. Consider what a great forest is set on fire by a small spark."* We must realize that our words can leave a wake of destruction and devastation behind. We should heed the advice of the old adage that says, "Put your brain in gear before you put your mouth in motion."

Romance, dating, and relationships are all part of the college life. But, what you listen to can kill you. Read Judges 16:17-22

Who owns the largest printing press in the world? The New York *Times*? Look? No! It is the Jehovah's Witnesses. They have one press that puts out 500 pieces of propaganda every second. From that one press alone, come out 84,000,000 books and pamphlets.

Many students are inundated with occulted materials when entering college. Some claim there is no Hell. Still others claim all good people go to Heaven. Be careful what you listen to! False doctrine can delay, distract, or destroy spiritual growth. Are you overcoming false doctrine today? Write down how you can propagate the truth in a way that will attract others. Is your life attractive to others for Jesus today? Write down why or why not.

Goals are important. Someone once said that if you aim at nothing, you will hit it every time. The words found in Matthew 25:21 represent the ultimate goal for most Christians, six simple words that could mean the difference between life or death eternal. What are they? *"Well done, good and faithful servant!"* Nothing could be sweeter than to hear those six words uttered by our Master.

In the fall of 1992, on *Prime Time Live*, Diane Sawyer interviewed the man many consider to be the greatest evangelist to ever live. Billy Graham is respected worldwide by everyone from kings and presidents to factory workers and peddlers. He said in the interview that he never felt as if he had done enough. And in his final statement, he spelled out his goal: "When I stand before the Lord, I just want to hear one simple sentence from Him. I want Him to say to me, 'Well done, thou good and faithful servant.' I don't know if He'll say it, but I hope He will." Let us long for those words more than money, fame, or success.

Sometimes it's easier to run from our problems than to stand and face them. Read Judges 7:7-16.

"I do the very best I know how—the very best I can; and I mean to keep doing so."

Abraham Lincoln's life is the best example of his own words. Consider the chronology of his career:

> 1831 – Failed in business
> 1832 – Defeated for legislature
> 1833 – Again failed in business
> 1834 – Elected to legislature
> 1835 – Sweetheart died
> 1836 – Had nervous breakdown
> 1838 – Defeated for speaker
> 1840 – Defeated for elector
> 1843 – Defeated for Congress
> 1846 – Elected to Congress
> 1848 – Defeated for Congress
> 1855 – Defeated for Senate
> 1856 – Defeated for Vice President
> 1858 – Defeated for Senate
> 1860 – ELECTED PRESIDENT

No matter how overwhelmed you feel today, God will make a way. Just as with Gideon, God has a strategy for your success today. Are you facing your problems or running away? Write down how you and God can overcome today!

Don't you love a story with a good ending? *Beauty and the Beast* is a classic story that was released on video by Disney in 1992. It tells a story of a prince who was put under a spell that turned him and his entire estate into an ugly sight. The spell could only be broken if he gave and received love. A beautiful girl named Belle fell in love with the Beast and the spell was broken. The movie ends with a joyous celebration.

Revelation 20:10 illustrates a spectacular ending that provides good reason for the Christian to celebrate. It reads, *"And the devil, who deceived them, was thrown into the lake of burning sulfur where the beast and the false prophet had been thrown. They will be tormented day and night forever and ever."* We are definitely on the winning side, and fortunately, this ending is no Disney classic.

JANUARY 26   NO! – IT'S MINE

By nature we are possessive. We call our cars ours. When we get a book we write our name in it. We want our possessions to be claimed as "mine." Read Ruth, chapter 1, verses 16-17.

One of the world's most renowned women was Helen Keller, that prodigy who lived and became famous—without sight or sound. But Helen Keller had another self, another half.

Anne Sullivan was born at Feeding Hills, Massachusetts, in poverty, in affliction. She was half blind. Her mother died and she went over the hill to the poor house. Then, at the Perkins Institute for the Blind, a brilliant operation restored her sight. Thereafter she devoted herself to the care of the blind.

Meanwhile, down south a baby was born, a girl destined after early childhood never to see or speak or hear! Helen Keller. She came under the care of Anne Sullivan. In two weeks Anne taught her thirty words, spelling them by touching the hand. Under this system, Helen Keller rose to renown. Teacher and pupil remained inseparable for forty-nine years.

Time came when misfortune befell Anne Sullivan, who meanwhile had become Mrs. Macy. What misfortune? She became blind. And now, turn about, Helen Keller taught her how to overcome the lack of sight. She schooled her former teacher as devotedly as she herself had been schooled.

Finally Helen Keller stood at the deathbed of her other half. When it was all over, she said: "I pray for strength to endure the silent dark until she smiles upon me again."

A drama of the dark!

God says we are his possessions. Does God possess all of you? Write down if you think God smiles on your life.

JANUARY 27   CHRISTIAN COPYCATS

Take a look at 3 John 11. It says, *"Dear friend, do not imitate what is evil but what is good. Anyone who does what is good is from God."* Rich Little is famous for imitating people. He has often been referred to as the man of a thousand voices, and is easily the greatest impersonator that has ever lived. Rich Little learns to imitate people by studying their habits, watching tapes of them over and over, and listening to how they talk. Children also learn by imitating the things they see adults do.

So how do we imitate what is good? We do so by surrounding ourselves with people who have Godly qualities and characteristics. We carefully choose the people we spend most of our time with, making sure they are people who will help us to become more like Christ. Whom are you imitating? Whose habits are you studying? What type of speech is coming from your mouth? Remember, imitate what is good!

JANUARY 28   MOTHER KNOWS BEST

Read Proverbs 1:8-9.

There are many situations in today's family. Sometimes our words can cut deeper than deep.

George Washington, when quite young, was about to go to sea as a midshipman. Everything was in readiness. His trunk had been taken on board the boat; and he went to bid his mother farewell, when he saw her eyes filling with tears. Seeing her distress, he turned to the servant, and said, "Go and tell them to fetch my trunk back. I will not go away to break my mother's heart." His mother, struck with his decision said to him, "George, God has promised to bless the children that honor their parents; and I believe he will bless you."

The Bible promises that if we obey our parents we will live a long life. When our family is hurting everything hurts. Are you blessing your family or cursing them today? Write down how you can bless your mom today. A card, a flower, or a kind word. Mother knows best.

JANUARY 29   A HANDYMAN'S SPECIAL

The newlyweds were thrilled when they found the advertisement for a handyman's special house in their price range. Remodeling would be fun and exciting. After a few weeks, they signed the deal and moved in. They immediately began pulling down old wallpaper, tearing down walls, and sanding doors and windows. The mess that was created was incredible. Dust covered everything. Trash piles were everywhere. It seemed like the place would never look nice and that they would never complete the task. But when that day finally arrived, the old house was a showplace.

Do you ever feel like that God is remodeling you? We ask God to make us a vessel of honor, a holy and righteous servant, but what is supposed to be progress feels like a big mess. As God begins tearing out "walls" of sin and rebuilding us inside, it feels as though the task will never be completed. Philippians 1:6 tell us, *"...that he who began a good work in you will carry it on to completion until the day Christ Jesus."* If we let Him, God will turn our lives into a showplace, a testimony of His grace and goodness.

Can you imagine really physically walking with God? Read Genesis 3:8-9.

Years ago a young missionary had to flee from western China. An infuriated mob hotly pursued him. He hastily boarded a river boat. The mob too came on the boat. Then he jumped into the river. The mob began to throw spears at him. Miraculously he escaped, unharmed. When he was telling of this ordeal later, a friend asked him, "What verse from the Bible came to you as you were darting beneath the boat to escape the spears of the mob?" "Verse?" He asked in astonishment, "Why the Lord Himself was there with me."

No matter what the world throws at you today, if you continue to walk with God, He promises to walk with you! Are you walking with God today? Write down ways to walk with God.

One day flannel is out, and then all of a sudden it is in. Tie–dye sweeps the 60's and then makes a comeback in the 90's. Do you wear a wide tie or a thin tie, Docker's or Bugle Boys, prep or grunge wear? Fashion keeps everyone guessing and everyone spending. Of the $57 billion that teens spent of their own money in 1992, much of it went to purchase clothes, and only the name brands like The Gap and Birkenstock will do. And let's face it, appearance is important.

Maybe that is why the Bible talks about first–class fashion in Romans 13:14. Do a little window shopping in this scripture and try this on for a striking appearance that will stand out above the rest: *"...clothe yourselves with the Lord Jesus Christ, and do not think about how to gratify the desires of the sinful nature."* If you "dress" your spiritual man with the world's fashion, you will look worn and haggard. But if you put on Christ, you will look like a million bucks!

Michael was craving Mom's cooking and looking forward to sleeping in his own bed. Finals were over, the car loaded, and he was on his way home for Christmas Break. Michael knew he was speeding, but he could almost smell his mom's cooking. The State Trooper, however, did not seem to understand and began writing a citation for speeding. "But sir," Michael said, "I had every intention of slowing down and going the speed limit."

Sadly enough, for many people the cravings and desire of life have put them in high gear with total disregard for God's laws. Someone once said that the road to hell is paved with good intentions. But good intentions are not enough. There are still consequences to disobedience. In Romans 8:7, 8 we read, *"the sinful mind is hostile to God. It does not submit to God's law, nor can it do so. Those controlled by the sinful nature cannot please God."* Turn your good intentions into good actions. Live in obedience and submission to the law of God.

Read 1 Samuel 17:45–50.

In today's generation, there are many giants you must face that past generations never had to face. Drugs, STDs, AIDS, the destruction of the nuclear family and economic uncertainty, only to name a few.

When the Emancipation Proclamation was taken to Mr. Lincoln by Secretary Seward, for the President's signature, Mr. Lincoln took a pen, dipped it in the ink, moved his hand to the place for the signature, held it a moment, then removed his hand and dropped the pen. After a little hesitation, he again took up the pen and went through the same movement as before.

Mr. Lincoln then turned to Mr. Seward and said: "I have been shaking hands since nine o'clock this morning, and my right arm is almost paralyzed. If my name ever goes into history, it will be for this act, and my whole soul is in it. If my hand trembles when I sign the Proclamation, all who examine the document hereafter will say, 'He hesitated.' "

He then turned to the table, took up the pen again and slowly, firmly wrote "Abraham Lincoln," with which the whole world is now familiar.

He then looked up, smiled, and said, "That will do it."

No matter what the giants may be on your campus, with determination and faith in God, you can conquer your giants and make you place in the history books of heaven. Write down your giants and how you can overcome them today.

FEBRUARY 3   AGREEMENT[2]

Read Matthew 18:19. It talks about agreement squared, what can happen when two people storm heaven with prayer. A minister and meteorologist played golf together every weekend. One Saturday, just after they teed off, a massive thunderstorm broke out. They ran back to the clubhouse, soaking wet, and waited for an hour. But the rain continued to fall. Finally, the minister turned to the meteorologist and said, "You would think that between the two of us, we could do something."

Well, according to God's Word, when the storm hits, we can do something. This scripture says, *"Again, I tell you that if two of you on earth agree about anything you ask for, it will be done for you by my Father in heaven."* When the pressure mounts, or life delivers a blow that really stings, find a prayer partner and exercise the power of agreement. Link your faith with the faith of a fellow warrior and do something about the situation.

Read Judges 2:18.

There are warning signs all around us – Slippery When Wet, Danger High Voltage, Reduce Speed Ahead, 55 M.P.H..

There is a huge painting hanging in the Supreme Court Building in the little country of Switzerland. It was painted by an artist named Paul Robert, and the title is "Justice Instruction the Judges."

In the foreground are the litigants—the wife against the husband, the architect against the builder, and the like. Above them stand the Swiss judges with their little white dickeys. How are these people going to judge the various litigations? A whole sociological theory is opened up.

The artist's answer is simply this: Justice (usually blindfolded, with her sword vertical as is common) is unblindfolded, with her sword pointing downward to a book on which is written "The Word of God."

There is law and warning all around us. To break the law can have life altering consequences. To break God's law will have eternal consequences. Are you obeying the Judge today? Slow down and write how you can obey the laws of the land and God's laws today.

The image will be imbedded in the minds of people world–wide for years to come. Young people fought for democracy in the nation of China, and the heated battle intensified to the point of death and bloodshed. As tanks rolled into Tienemen Square in a single file line, one lone teen stood in the path of the tanks and defied their progress.

We should have the same attitude as the enemy of our soul tries to take over and rob us of life and freedom. We must stand our ground and not bow to the threat or onslaught of Satan. Paul says in 1 Corinthians 15:58, *"Therefore, my dear brothers, stand firm. Let nothing move you. Always give yourselves fully to the work of the Lord, because you know that your labor in the Lord is not in vain."* Just as worthwhile and heroic as the young teen's efforts to fight for democracy are our efforts to uphold God's standards. Don't be moved by adversity, but be rock–solid in God's work.

FEBRUARY 6   WHAT WOULD DADDY DO?

Read Psalm 91:1-16.

Father, Daddy always seems to have the best advice. In careers, choices for the future, financial matters, it's always nice to call Dad and say, "What would you do?"

In ancient Israel six cities were founded as cities of refuge. Thither for refuge could flee men who, without malice or premeditation, had taken the life of a fellow man. Once within the gates of the city of refuge, they could not be touched by any hand of vengeance or judgment. The rabbis have an interesting tradition that once every year the roads leading to these cities of refuge were carefully repaired and cleared of obstacles and stones, so that the man fleeing for his life would have no hindrance in his way. The Cross is God's great and eternal city of refuge from the penalty upon sin.

Our heavenly Father has provided safe passage for you today. In all the confusion and loneliness of college life, God's word, God's church and God's people provide safe passage. Your earthly father may disappoint you but God, your heavenly Father, will never disappoint you. Write down how Daddy God can help you today and send a note to your Dad! What would Daddy do?

FEBRUARY 7   LAST PLACE IS FOR LOSERS

So goes the message of the world day after day, from Little League baseball games to the cover of *Fortune* magazine. We are constantly reminded that losers come in last, quite the opposite of what Jesus said in Mark 9:35. *"Sitting down, Jesus called the twelve and said, 'If anyone wants to be first, he must be the very last, and the servant of all.'"* When we practice these words, we discover what it means to be a true winner, and God promotes us to the winner's circle.

Take Bob Wieland for example, a man who lost both legs when he stepped on a land mine in Vietnam. He finished the New York Marathon in 1986 dead last, behind 19,412 competitors. It took him 4 days, 2 hours, and 17 minutes. Wieland told the L.A. Times in a front-page story, "I'm a born-again Christian and this was a demonstration that faith in the Lord Jesus will always overcome the impossible." A true winner!

FEBRUARY 8   PIMPLE FACED AND SHORT TOO!

King David as a boy, the Bible says, was red faced "ruddy" or pimpled and short. Not a dynamic description of a king. Right?

COLUMBUS was the son of a weaver, and a weaver himself. CERVANTES was a common soldier. HOMER was the son of a small farmer. MOLIERE was the son of a cutler. TERENCE was a slave. OLIVER CROMWELL was the son of a London brewer. HOWARD was an apprentice to a grocer. FRANKLIN was a journeyman printer, and son of a tallow-chandler and soap-boiler. DR. THOMAS, Bishop of Worcester, was the son of a linen draper. DANIEL DEFOE was a hostler, and son of a butcher. WHITEFIELD was the son of an innkeeper at Gloucester. VIRGIL was the son of a porter. HORACE was the son of a shopkeeper. SHAKESPEARE was the son of a wood-stapler. MILTON was the son of a money-scrivener. ROBERT BURNS was a ploughman in Ayrshire. MOHAMMED, called the prophet, was a driver of asses. MOHAMET ALI was a barber. MADAME BERNADOTTE was a washerwoman of Paris. NAPOLEAN, descendant of an obscure family of Corsica, was a major when he married Josephine, the daughter of a tobacconist Creole of Martinique. GEN. ESCARTERO was a vestry clerk. BOLIVAR was a druggist. VASCO Da GAMA was a sailor. JOHN JACOB ASTOR once sold apples in the streets of New York. CATHERINE, Empress of Russia, was a camp-grisette. CINCINNATUS was plouging in his vineyard when the dictatorship of Rome was offered him. ELIHU BURRITT was a blacksmith. ABRAHAM LINCOLN was a rail splitter. GEN. GRANT was a tanner. COM. VANDERBILT was a ferryman. DANIEL DREW was a poor widow's son. SENATOR WILSON was a cobbler. GEN. BANKS says that he graduated at a university which had a waterwheel at the bottom, and a bell at the top.

Today, you are in college, many times, feeling inadequate and inferior. You are in position for greatness and only you hold the key! It's not what you've done, or where you've been, but Whose you are! Do you belong to Jesus? He can make you great. Write down how God can advance you today!

FEBRUARY 9 $A + B = C$

Oh, those ungodly Algebra equations. There you are, studying numbers, and they keep throwing in these crazy letters. It seems there is an equation or formula for everything, and all the while, you are thinking, "I'm never going to use this stuff!" "But it teaches you problem solving," states the professor, not realizing that he is a part of your problem.

God's Word has some very practical equations. But fear not, they are easy to understand and are useful. James 4:7 offers us such an equation, one that we should apply to our lives every second of every day. *"Submit yourselves, then, to God. Resist the devil, and he will flee from you."* First, we must submit (A), then resist (B). These two added together means the devil will flee (C). Simple enough. And it works! The result is a life that is in God's hands, and total victory over our enemy.

Read Exodus 20:14-17.

Adultery and fornication are trivialized in today's liberal media. Soaps, sitcoms, movies, and videos all make it seem as natural as brushing your teeth to run around on your mate.

A Greek student in Athens, Greece, was sentenced to eight months imprisonment on charges of marrying two women within 48 hours. He appealed the sentence and was set free pending a new trial. The court heard that Petros Novaras, 29, married Vassiliki Chioti on January 24, 1971 in the central Greek town of Lamia, and took off in his car for a honeymoon. After an engine trouble, he sent his wife down to Athens on a bus.

In the meantime, he went to a suburb in Athens and married a 29-year-old lass, the court heard. He then continued his honeymoon with his second wife. At the court the accused testified, "Both families were putting unbearable pressure on me. So I decided to take them both so as not to hurt anybody's feelings."

God's word is clear that faithfulness in marital and single relationships is commanded. No matter how liberal America's colleges become, abstinence in sexuality is God's policy. Write down how you can live in purity for God in your sexuality.

The invention of the Walkman radically changed a very precious commodity in the lives of individuals—solitude. No longer would a person have to be without noise or "companionship" ever again. Most people are in the habit of turning on the stereo or TV as soon as you walk into your room or apartment. But now, with the Walkman, you can have someone talking to you or singing in your ear wherever you go. But we miss something when we forfeit quiet times and places of solitude.

In our busy world, with crowded schedules and an array of activities, it takes a conscious effort to be alone with God. Times of quietness alone with God are essential for us, for without them our spirit dies. Jesus himself made the secret place a priority, as we see in Mark 1:35. It reads, *"Very early in the morning, while it was still dark, Jesus got up, left the house and went off to a solitary place, where he prayed."* There is something about starting our day off in prayer, alone with God, that insures a successful relationship with the Lord. Make a place in your daily schedule to spend time in the secret place.

FEBRUARY 12   ONE MORE FOR THE ROAD

Read Leviticus 20:26

Alcohol consumption is almost synonymous with college life. 90% of America's college students say they drink at least occasionally.

The National Institute on Alcohol Abuse and Alcoholism says that problem drinkers are those who may not yet be full-fledged alcoholics, but whose drinking, seriously and adversely, affects their lives. Symptoms:

- Drinking regularly in order to function or to "cope" with life.
- Drinking to get drunk frequently.
- Going to work intoxicated.
- Driving a car while drunk.
- Requiring medical attention because of drinking.
- Getting into trouble with the law while drinking.
- Doing something while drunk which a person would never do sober.

God asks for us to be holy as He is holy. Holy means "set apart for the use of God." Are you being drowned in the tide of alcoholism or alcohol abuse? Every alcoholic started with one for the road! Write down how you can be holy for God today.

The commercial showed a group of business men and women crammed into a VW, carpooling to work. The announcer then makes this statement, "Aren't you glad you use Dial? Don't you wish everybody did?" So, here's the question for you: How do you smell? Okay, let's assume you practice good hygiene. But, how does your life smell spiritually?

If you have done much driving in the country, chances are you have caught a whiff of a skunk, which can be smelled for miles. Up close the smell can be nauseating. Well, the same is true for our lives. If we are out of fellowship with God, or we are living a lie, we become an unpleasant odor to this world. On the other hand, when we are in communion with God, our lives are a sweet smell, a breath of fresh air to the world. Read 2 Corinthians 2:14. *"For we are to God the aroma of Christ among those who are being saved and those who are perishing."* May our lives be the sweet aroma of Christ to a world that is full of filth.

God makes it clear to us in His Word that because He has extended love to us, we are to love one another. In fact, He even tells us to love our enemies. Society teaches that love is a feeling when in fact, love is a choice of one's will to honor, accept, and show kindness to another person. Good feelings come when we obey God's Word. Loving someone goes far beyond merely tolerating them. It requires action, a demonstration of honor, acceptance, and kindness.

A female student from the University of Southern California decided to distribute Valentine cards to her friends. Not wanting to create more tension in an already-strained relationship with her roommate, Kim delivered a card to her roomy's mailbox with a slight correction. She had carefully crossed out "I love you" and printed "I don't mind you" instead. 1 Corinthians 13:13 says, *"And now these three remain: Faith, hope and love. But the greatest of these is love."*

Read II Kings 6:15-17.

Countless accounts abound where men and women would bet their life that some kind of angel assisted them in a crucial situation.

In *A Slow and Certain Light*, Elizabeth Elliot told about her father's experiences with angelic helpers:

"My father, when he was a small boy, was climbing on an upper story of a house that was being built. He walked to the end of a board that was not nailed at the other end, and slowly it began to tip. He knew that he was doomed, but inexplicably the board began to tip the other way, as though a hand had pushed it down again. He always wondered if it was an angel's hand.

The scripture declares that in the spirit realm angelic hosts are all around the righteous to protect and provide God's blessings and benefits. In times of crisis or loneliness, call out to God to know that God's angels are all around you—even chariots of fire. Write down how God has protected you recently. Who knows? It might have been an angel!

Read Numbers 11:1.

With baby boomers grown up and baby busters stepping into the scene, America's youth and young adults are more angry than ever. George Barna, in his book *What Americans Believe*, found 67% of Americas college students are angry. That America will not get better, but in fact, will get worse and even the American dream is in jeopardy.

The great Maestro, Toscanini, was as well–known for his ferocious temper as for his outstanding musicianship. When members of his orchestra played badly he would pick up anything in sight and hurl it to the floor.

During one rehearsal a flat note caused the genius to grab his valuable watch and smash it beyond repair.

Shortly afterward, he received from his devoted musicians a luxurious velvetlined box containing two watches, one a beautiful gold timepiece, the other a cheap one on which was inscribed, "For rehearsals only."

Anger can cause us to say and do things we later regret. What makes you angry? Ask God to help you use your anger to help not hurt. Write down how you can overcome anger today.

FEBRUARY 17   ALWAYS AND FOREVER

Heraclitus once said, "There is nothing permanent except change." Things are constantly changing around us. Just when you finally invest in a new car, they change the body style. You finally decide to purchase a CD player, and they introduce a slightly different but better component. The stock market changes daily, along with the weather, taxes, athletes salaries, interest rates, the top–40 list, and public opinion.

Heraclitus was right with one exception. Two things are permanent, change and Jesus Christ. In the midst of whirlwind changes that will cause you head to spin, the Bible teaches the permanency and consistency of Christ, traits that bring stability to our lives. In Hebrews 13:8, we find the comfort of these words: *"Jesus Christ is the same yesterday and today and forever."* His love never changes. His faithfulness and mercy never change. His power and ability to make a difference in our lives is the same always and forever.

FEBRUARY 18   MIRROR, MIRROR ON THE WALL

Read Deuteronomy 6:4-9.

The telephone book in its church and religious section abounds with hundred of categories–Baptist, Pentecostal, Methodist, Catholic, Mormon, Islam, Buddhist, etc. Each professes knowledge of the one true God. Do not be deceived—God is not mocked.

Almost 50 years ago, when Dr. Mosinsohn of the Hebrew College of Jaffa was touring America, he remarked at the University of California:

"Think of all the great religious leaders who have come out of the East. Moses arose in the East; Buddha, Confucius, Jesus and Mahomet all arose in the East. And we say to you people of the West, with confidence, that if you will restore the Jew to his ancestral home, it will not be long until we give you another great religious leader who will perhaps transcend all who have gone before."

The Anti-Christ will arise soon most think from the Eastern nations. In college dorms, class rooms and lecture halls, God is minimized to your own mind and New Age humanistic philosophy. God's Word is clear—there is only one God. Have you asked this one true Jehovah God into your life? Write down how God is real in your life.

FEBRUARY 19   HOME ALONE

Do you remember what it was like as a small child to suddenly discover you were all alone? Many children live in fear of being abandoned. That fear was a reality for Nicole and Diane Schoo. In December of 1992, their parents left these two young girls, ages 10 and 4, at home while they went to Mexico on a nine—day vacation.

Maybe now, as an adult, you have good reason to fear being left alone. Maybe you have brought fears from your childhood into your adult life, and now find it hard to trust anyone, even God. Rest assured, our Heavenly Father will never abandon us. In Hebrews 13:5, we read, *"...God has said, 'Never will I leave you; never will I forsake you.'"* That is a promise from God, and we can trust Him today with full confidence that He will never leave us alone.

Read Joshua 1:8-9.

Fear is the number one reason people do not share their faith in Christ. Fear of rejection, fear of not knowing what to say and fear of losing or tainting their reputation.

Here's something that sounds like a joke, but it's a tragedy. It happened in Spain. In Barcelona a truck was rolling along, carrying an empty coffin. A farmer who was hitch–hiking thumbed a ride. He was bouncing along in the rear of the truck, which was open, when it started to rain. He examined the coffin, found it empty, and crawled inside to keep dry. There he fell asleep.

Further on, two other hitch–hikers got a ride on the truck. They were going along at a lively clip when the farmer inside the coffin pushed open the lid, stuck his head out, and observed: " Oh, it has stopped raining." The two other hitch–hikers were so terrified that they leaped from the speeding truck. One was killed.

Fear can cripple you in your walk with Christ. God told Joshua, I will be with you. God has promised to be with you today on your campus, in your room or walking to class.

Write down ways you can overcome fear with God's presence.

Wherever the President of the United States goes, he is surrounded by a squad of bodyguards known as the Secret Service. Their responsibility is to stay alert at all times and to protect the President from any bodily harm. Imagine the security a person would feel with that kind of full–time protection.

As Christians, we should feel the same sense of security. God has assigned secret service agents to us to constantly protect us from any harm or danger. They are known as angels. Many people can relate stories and accounts of how supernatural intervention took place in their life to protect them from death or danger. One man remembers how a "man" walked up to him as he sat crying in the middle of the woods. He was completely lost, the sun had gone down, and there was no response to his gunfire cry for help. The "man" walked up out of nowhere and gave him specific directions to get out of the woods. In Luke 4:10, God's Word tells us that *"He will command his angels concerning you to guard you carefully."*

FEBRUARY 22   GOD IS NOT DEAD

Read 1 Samuel 2:9-10.

Many people today reject the thought of a living, omnipotent, omnipresent God who transcends time and space. However, this does not do away God.

Madalyn (Murray) O'Hair, the atheist who brought about the Supreme Court decision on school prayer, filed suit in Federal Court against President Nixon, the treasurer of the U.S., the Senate and House chaplains and other Congressional officials for allowing services in the White House and Capitol.

Mrs. O'Hair, who acted as her own attorney, accused Mr. Nixon of being the "central figure" in an effort to "make Christianity the official 'civil religion' of the United States." She specifically charged the President with holding religious services in the White House in violation of the first amendment.

In her suit, she asked the court to enjoin Mr. Nixon from allowing such services and to declare unconstitutional the practice of "devoting the property and premises of the Executive Mansion" to such religious services.

You can oppose the President and live, but anyone who opposes God in defiant rebellion will be doomed for destruction. Are you opposing God in areas of your life? Write down how you can walk in obedience.

Are you a person of honesty and integrity even when no one is looking? We are all tempted to get away with certain things when no one will notice, like cheating on an exam or "borrowing" a room mates AT&T calling card. Raymond Floyd, a professional golfer, was among the leaders in a tournament where the 1st place prize was over $100,000. As he prepared to tap in a routine 9–inch putt, he accidently touched the ball with his putter. According to PGA rules, if the ball moves at all, a stroke must be counted, which in this case could cost Floyd a lot of money. What would he do?

In Romans 12:17, we read *"Be careful to do what is right in the eyes of everybody."* God will always honor us and bless us for being honest. It may cost us some temporary gain to do what is right, but the lasting rewards far outweigh what is lost. Floyd could have looked around to see if anyone was looking, or acted like a bee swarmed his head. But he did what was right, assessed himself a stroke, and took a bogey, even though he barely touched it.

Read Genesis 3:1-15. Have you ever been poisoned?

On March 13, 1967, a jet plane, flown by an Air Force pilot, released 320 gallons of "nerve gas" that had been developed by the United States. A sudden shift in winds caused less than 5% of the released load to fall, accidently, 20–30 miles from the planned clear target. This small amount of nerve gas settled on thousands of grazing sheep killing, almost instantly, over 6,000 sheep.

Poison can be a liquid, gas or solid or it can be words. Satan poisoned Eve with eating a deadly potion–fruit from the Tree of Life. Are you being poisoned today? Is there something coming into your life that God has warned you not to do? Write down areas of your life that are being poisoned by what you hear, see, say, touch or by places that you go.

Over the last ten years, the number of teens who are using anabolic steroids has increased at alarming rates. Several studies indicate that one of every 15 male high–school seniors are using steroids. People in the 90's want to get "totally pumped up" and are willing to pay the price to have the perfect physique. Health spas and weight gyms are flooded every day by people who are building their bodies. But how much time do we spend building the spiritual man?

Jude 20 says to *"build yourselves up in your most holy faith..."* So here in verse 20 and 21 is offered a great workout program for the Christian.

1. Read and study God's Word (the holy faith) daily.
2. Pray in the Spirit. Allow the power of the Holy Spirit to help you pray more effectively.
3. Remain in God's sphere of love.
4. Long for, prepare for, and anticipate God's return.

Let's get pumped up for God!

Mathematics can sometimes be a big problem for many people. Read Psalms 1:1-6.

Life is a book of Volumes Three. The Past, the Present and the Yet–To–Be. The first is written and laid away. The second we are writing day by day. The next and the last of Volumes Three is locked from sight–God holds the key.

Many things in life are a process–building a relationship, getting a diploma. In Psalm 1, we see this process of life can bring blessing or curses! Which process are you developing? A life of God's blessings or curses? Write down how you can process your life with God.

The young Marine lay in a hospital bed, clinging to life, his body mangled by the explosion from a grenade. Because of the young Marine's heroic war efforts, one of the top commanding generals decided to pay him a visit. As he stood by his hospital bed, he noticed the Marine trying to say something. He leaned down so that his ear was inches from the man's mouth. He then heard the young Marine whisper the words "Semper Fi", the two words inscribed on the Marine logo which mean "always faithful."

Should our commitment and attitude as a Christian be any less? Even in the midst of trials, persecution, or even death, we should be "always faithful" to our Commander in Chief, God Himself. Revelations 2:10 tells us that those who do so will be rewarded eternally. It says *"Be faithful, even to the point of death, and I will give you the crown of life."*

Do you remember playing this game as a kid?
Read Joshua 7:10-15,24-26.
Achan sinned and tried to hide it. He and his whole family paid a severe penalty. Others had to pay because of his sin. God does not want us to hurt the spiritual lives of other Christians. Our sins can cause harm not only to us but to others—possibly even to our own family.

Have your actions ever caused someone to question God? Are you leading someone astray now? Ask God to forgive you, then go to that person as quickly as possible and ask forgiveness.

When we hide, God will seek us out! Repent of unconfessed sin today and let God cleanse you.

"I just have a hard time saying 'no'," has gotten more people in trouble than any other excuse known to man. Could you say no to a million dollars? That was the question posed to Demi Moore by Robert Redford in the movie *Indecent Proposal.* Redford offers Demi's husband in the movie one million dollars if he will let his wife have sex with Redford. Many Americans surveyed said they would take the offer. Thus, the ethics of Americans is challenged once again.

At some point, we must learn to say no to sin, regardless of how fun or gratifying (temporarily) it may be. And God's Word tells us that the grace of God, after we are saved, teaches us to do just that. Of course, we ultimately must utter the words and walk away from temptation. Titus 2:12, speaking of God's grace, says, " *It teaches us to say 'NO' to ungodliness and worldly passions, and to live self—controlled, upright and godly lives in this present age."*

Read 1 Samuel 14:1-15.

It has been said that you will do things with others that you would never do alone.

Napoleon had in his school at Brienne a young friend, Demasis, who greatly admired him. After Napoleon had quelled the mob in Paris and served at Toulon his authority was taken from him, and he was cast out penniless. He even meditated suicide, and was on his way toward the bridge from which he expected to throw himself, when his friend, Demasis, met him and asked what was the matter.

Napoleon frankly told him he was without money, his mother was in want, and he had despaired. "Oh, if that is all," said Demasis, "take this; it will supply your wants," and he handed him $600 in gold.

Napoleon said afterward that he hardly knew why he took it, but he did, and rushed off to his cottage home. When Napoleon came to power he sought for Demasis far and wide. He wanted to promote him, he wished to enrich him, and it was said that Demasis lived and served in one of his armies, but would not make himself known.

Jonathan had a friend who encouraged and believed in Him. Do your friends in college inspire you to greatness or drag you to defeat? Write down how you can be a friend and how your friends can help you to live for Jesus as did Jonathan and his armor-bearer!

MARCH 3 *STEP UP TO THE LINE AND GO FOR IT!*

Whenever we approach God with our needs, He desires that we do so with confidence, with expectancy. In fact, in 1 John 5:14-15, we read, *"This is the confidence we have in approaching God: that if we know that he hears us–whatever we ask–we know that we have what we asked of him."*

Don Calhoun knows about confidence. He stepped up to the free throw line at Chicago Stadium April 14, 1993 with confidence. The Bulls were playing the Miami Heat and Don, a 23 year old office supply salesman, was selected for a once–in–a–lifetime opportunity. If he hit one shot from three–quarters of the length of the court, he would win one million dollars, and he was only given one try. One was all it took. Calhoun swished the shot and became an instant millionaire. "I stepped to the line with lots of confidence," said Calhoun.

God desires that we approach Him with the same type of confidence, knowing that God hears us and will answer us, and that we will leave His presence winners.

Read 2 Samuel 12:5-7.

When confronted with a hard fact about ourselves, often it's hard to humble ourselves and admit that weakness exists.

Pearl Harbor! The morning of December 7, 1941 found 353 Japanese airplanes swarming all over the Harbor site. Within a couple of hours, America lost 8 big battleships, 6 major airfields, almost all planes, and 2,400 men. That happened at 7:50 am in what was supposedly a surprised attack. But these are the startling facts.

That morning at 7:00, while the Japanese warplanes were 137 miles (50 minutes) away, two US soldiers on a small radar station in the Pacific scanned the screen and saw dots and dots appearing, until the whole screen was filled. These soldiers notified their youthful supervisor, a lieutenant. No other officer was around, that being a Sunday.

The lieutenant thought these must be planes from California, and without another thought, said these crucial words: "Don't worry about it." There would have been time to scramble the planes at Pearl Harbor, prepare the battleships and shelter the men, but this lieutenant, at the most responsible moment of his career, failed the nation.

David was confronted and admitted his defeat and God restored him, even calling him a man after God's own heart. Has someone confronted you? If you do not correct this, destruction can come to you just like Pearl Harbor. Write down how you can correct your weaknesses.

MARCH 5   RESCUE ME, IN YOUR LOVING ARMS

Can you imagine what it would feel like to be rescued from cold, shark–infested ocean waters? Walter Wyatt, Jr. can. It happened to him in December of 1986 after his twin–engine plane went down between Nassau and Miami. Wyatt floated on a life vest for more than 15 hours, while sharks circled and threatened to have him for lunch. At 9:00 am, he was plucked from the sea by a Coast Guard cutter, rescued from danger.

Maybe your story isn't as dramatic, but if you have been redeemed by Christ, you have been rescued. In our darkest hour, with sharks of sin and death circling around us, God plucked us out and rescued us from danger. Read Colossians 1:13-14.

MARCH 6   ACCUSATIONS

Read Ezra 4:6, 11-24.

Have you ever been falsely accused? The family of a man wrongly accused of shoplifting was awarded $107,000 by a jury which was convinced that his death at age 53 was caused by a broken heart. The charge was shoplifting a 63-cent can of Danish bacon in a Sacramento, CA drug store. John F. Abercrombie was a retired Air Force colonel with a distinguished World War II combat record. Abercrombie, also a scoutmaster, was employed as a civilian analyst for the California Highway Patrol. He was found innocent of the shoplifting charges, but his whole life was tainted. He lost his zest for life, became depressed and died of a heart attack shortly after the trial.

Satan, our accuser, wants to destroy our lives as well. Our God convicts not condemns, accepts and does not accuse. Are you falsely accused or are you accusing someone else falsely? If so, write down those accusations and let God bring you healing today.

MARCH 7   HOW'S YOUR ATTITUDE THESE DAYS?

In April of 1993, the news media made celebrities of a group of high school guys from California known as the "Spur Posse." What did this gang of youths do to gain such attention nation–wide? They competed with each other to see who could have sex with the most girls, and even designed a point system to tally their "conquests," one of whom was an 11-year-old girl. When interviewed on shows like ABC's 20/20, the gang members displayed some of the most arrogant, calloused, prideful attitudes the nation has ever seen in a young person. They obviously needed an attitude check.

We find in Philippians 2:5-11 the perfect attitude checklist to keep ourselves Christ–like. Verse 5 says, *"Your attitude should be the same as that of Christ Jesus."* It goes on in the following verses to tell us what that is. It speaks of humility, servanthood, and obedience. Read it and take a self–inventory of your attitude.

MARCH 8   QUITTERS NEVER WIN

Read 2 Kings 4:28-35.

You have probably heard quitters never win and winners never quit. Winners are former losers who got tired of losing. You can never get down the road until you start the car.

During the pastorate of Henry Ward Beecher in Indianapolis he preached a series of sermons upon drunkenness and gambling, incidentally scoring the men of the community who profited by these sins. During the ensuing week he was accosted on the street by a would–be assailant, pistol in hand, who demanded a retraction of some utterance of the preceding Sunday.

"Take it back, right here!" he demanded with an oath, or "I will shoot you on the spot!"

"Shoot away!" was the preacher's response as he walked calmly away, hurling over his shoulder this parting remark:

"I don't believe you can hit the mark as well as I did!"

If you pray for something and it does not happen, know that God always answers prayer with yes, no or wait. Keep praying and believing and God will provide the way for you. There are two ways to climb an oak tree–you can climb it or sit on an acorn.

Write down how you can win with God and climb to the top. Quitters never win.

Can you imagine what life would be like without names? What if everyone were a number, or if cities were only identified by map coordinates? Names give identity and often reflect certain characteristics of a person or place like Happy, Texas–yes, there is a town with such a name. Maybe you would like to visit Tranquility, New Jersey or spend some time in Pleasureville, Kentucky. Chances are that not everyone in Happy, Texas is always happy. There are probably some people in Tranquility who are not really tranquil.

People who have accepted Christ into their lives as Lord and Savor are referred to as Christians. In fact, in Acts 11:26 it says, *"The disciples were called Christians first at Antioch."* The word "Christian" simply means Christ–like. It is important that the name "Christian" be reflective of our Christ–like characteristics and traits. If such is not the case in your life, begin today making that a daily goal and a top priority. It is a name worth living up to!

Do you have friends who encourage you to walk with God? Read 1 Samuel 14:1-15.

A Jesuit theologian declared in New York that after 400 years Anglican and Roman Catholic scholars finally sat down together to study their church's ordained ministers and found they had no basis for disagreement on the doctrine of the ministry. "It's astonishing, we believe exactly the same thing," said Father Herbert Ryan, S.J. Professor of Historical Theology at Woodstock College and Union Theological Seminary.

1 Samuel tells of Jonathan and his armor–bearer who believed in him. In verse 7 of chapter 14, his friend said, "go ahead, I am with you heart and soul." What a friend. You will do things with others you would never do alone! There is strength in numbers! Do your friends encourage you to do God's will or your own? Write down which way the "numbers" push you.

Children have a way of making things seem so simple. We adults like to do the opposite. We want to make things complicated. Maybe that is why Jesus said in Luke 18:17, *"I tell you the truth, anyone who will not receive the kingdom of God like a little child will never enter it."* There are many things about the Christian life that we must simply accept by faith.

After all, the Gospel is really very simple. A teacher was describing the Statue of Liberty to her second graders and pointed out that Lady Liberty had a torch in one hand and a book in the other. When she asked the students why the statue held a torch, one boy raised his hand and answered, "Because you're not supposed to read in the dark." Let's start acting like children when it comes to God's Kingdom.

Do you remember this song? "Take one down, pass it around, 98 million to go to the end of the song." It seemed eternal, and almost impossible to ever finish.

At a moderate rate, the Bible can be read cover to cover in about 90 hours or 15 minutes a day. The entire New Testament can be read in 20 hours, which is only 3 minutes a day for a year. Although it takes such a small percentage of one's total time, very few Christians will read through the Bible...in their entire lifetime!

Read Psalm 119:97-105.

This Psalm is the longest chapter in the Bible–176 verses. Its subject? The importance and blessings of reading, studying, memorizing and practicing the Word of God! What do you think God is trying to tell us? How much time each day are you willing to give to reading His Word?

Write down the verses you know, then begin to read through the Bible–15 minutes a day.

Life is full of temptations! At times, it seems impossible to resist or to overcome. But, what encouragement we find in 1 Corinthians 10:13. It is a promise to believers, one you can stand on. *"No temptation has seized you except what is common to man. And God is faithful; he will not let you be tempted beyond what you can bear. But when you are tempted, he will also provide a way out so that you can stand up under it."*

God will *always* do his part to help us in our time of weakness. We must do our part of resisting temptation. Then, we will escape peril. God provided an escape for Daniel from the lions, for the Israelites from Pharaoh, for Shadrach, Meshach, and Abednego from the furnace, and for Jesus from the tomb. He *will* provide a way out for us. Count on it!

What if a car pulls into a service station to fuel up–but never stops? It just keeps circling the pumps while the station attendant struggles to remove the gas cap. Finally, he just points the gasoline nozzle in the direction of the car and tries to get in as much as he can! Of course, the car is burning more fuel than it is receiving.

That's how we are sometimes. We may live at such a fast pace that we burn off more than we allow God to put back in us.

Notice how God spoke to Elijah–in a still, small voice. Read 1 Kings 19:1-12.

Will you slow down enough today to give God a few extra minutes to fuel you spiritually? If you're like many, you may be almost on empty. The world says to rush. Remember–if the devil can't make you bad, he'll make you busy!

Stop now and talk to God. He really wants to talk to you!

Someone once said that a friend is someone who is there with you when he would rather be anywhere else is the world. How far would you go for your friends? Four men in Mark 2:4 show us what true friendship is all about. They had a friend who was a paralytic and were determined to get him to Jesus while Jesus was in Capernaum.

"Since they could not get him to Jesus because of the crowd, they made an opening in the roof above Jesus and, after digging through it, lowered the mat the paralyzed man was lying on." These four friends went the extra mile in order to help their friend. Jesus admired their faith and diligence, forgave their friend of his sins, and then miraculously healed him. Will we do whatever it takes to get our friends to Jesus? That is what friends are for!

The African impala (that's an animal, not a car) can jump to a height of over ten feet and cover a distance greater than thirty feet in a single bound. Yet, these incredible animals can be kept in any zoo inside a three foot wall. Sound crazy? It's true. Why? Because impalas will not jump if they can't see where their feet will land.

Are you like the impala? Are you allowing doubt to entrap you in some kind of flimsy enclosure? Faith is the ability to trust in what we cannot see, and fear can hinder faith.

Read Isaiah 43:1-2.

Do you have the faith to take these verses at face value as words promised to us by God Himself? If you do, then you have nothing to fear.

What is the three foot wall in your life? Write it down. As you talk to God today, ask Him to help you have the faith to jump over it, certain that you will land on His promise to be with you. Don't be like the impala. You are never standing on a more secure place than when you are standing on God's promises.

A husband came home one evening and asked his wife what they were having for supper. "Take out," she replied.

"What kind of take–out?" he asked.

"Me!" she replied. Some people call it cabin fever, but if you stay cooped up for too long in one place, you begin to grow fidgety and uncomfortable. The same is true for the Christian. If we stay closed up inside the church walls too long, we become stagnant in our walk with the Lord. God has called us to go out and compel people to come in. We must share our faith.

God gives us our "take–out" orders in Luke 14:23. We ,must take our witness and our faith to people, and not sit back and expect them to come in. Jesus said, *"Go out to the roads and country lanes and make them come in, so that my house will be full."*

The Mercury and Gemini projects were the pioneer space excursions for the United States. The general manager of the projects was Walter F. Burke, who also taught Sunday School in his church. In an interview he affirmed, "I have found nothing in science or space exploration to compel me to throw away my Bible or reject my Savior, Jesus Christ, in whom I trust. The space age has been a factor in the deepening of my own spiritual life. I read the Bible more now. I get from the Bible what I cannot get from science–the really important things of life."

Read Genesis 1:14-18.

Men have come up with many theories to attempt to explain creation, while God has explained it in detail in His Word. The Bible is the ultimate physical science and biology textbook, because it states the origin of all things! Several theories are taught today. Pray that you will always remember the true Creator.

Have you ever made something hard out of something that is really very simple? How often do people do that with the Christian life and the message of salvation? The Gospel is really very simple, yet we often overlook the obvious. 2 Corinthians 4:4 says, *"The god of this age has blinded the minds of unbelievers, so that they cannot see the light of the gospel of the glory of Christ..."* Satan is the author of confusion and causes so many people with good intentions to miss the obvious.

That happened in 1974 to the Consumer Product Safety Commission, which brought out 80,000 buttons to promote its campaign for safe toys. The buttons read "For Kids Sake, Think Toy Safety." All 80,000 buttons had to be recalled. They were found to be unsafe. They had sharp edges and dangerous lead paint. Let's not overlook the obvious!

Read 1 Kings 9:6-7.

When we see a police car, the immediate reaction is to check our speed, seat belt, mirror and other mechanical responsibilities. To conform to the law can save a lot of money and lives.

A man starting in the fish business hung out a sign, "Fresh Fish for Sale Today," and invited his friends to the opening. They all congratulated him on his enterprise, but one suggested his sign might be improved. Said he, "Why the 'Today'? Of course it's today, not yesterday or tomorrow." So the fishmonger removed the word. Another said, "Why the 'For Sale'? Everybody knows that, else why the store?" And off came the word. Another complained, "Why the word "Fresh'? Your integrity guarantees every fish to be fresh." Finally, only 'Fish' remained, but an objector said, "Why the sign? I smelled your fish two blocks away!"

Conforming to God's laws brings life and disobedience brings death, sometimes to things very important to us such as family friendships, virginity or your GPA. Are you conforming to God's laws or your own? Burger King's latest slogan is wrong–"Breaking the Rules" can be devastating.

Write down how you can conform to God's laws.

MARCH 21   BE CAREFUL, LITTLE EARS, WHAT YOU HEAR

The Bible warns us many times to be careful not to follow the teachings of false prophets. In fact, 2 Peter 2:1 says, *"But there were also false prophets among the people, just as there will be false teachers among you. They will secretly introduce destructive heresies, even denying the sovereign Lord..."* and is God's Word ever true!

On April 23, 1992, on a national daytime television talk show, a woman discussed her role as the high priestess of sex in the church of the Most High Goddess, where sex is sacred. She claimed to have sex for religious purposes, to purify people. In March of 1993, a man who claimed to be Jesus Christ held cult members captive and federal agents at bay in Waco, Texas. False prophets and teachers abound, and many are not as obvious as these. We must guard our hearts *and* our ears, and pray each day that God will give us the gift of discernment.

MARCH 22   SOARING OR FLOPPING?

Do you sometimes wonder if you'll ever be able to soar like an eagle? C.S. Lewis wrote, "It might be hard for an egg to turn into a bird; it would be a jolly sight harder for it to learn to fly while remaining an egg. We are like eggs at present. And you cannot go on indefinitely being just an ordinary, decent egg. We must be hatched or go bad."

Are you still inside the shell thinking it would be impossible for you to fly? Read Isaiah 40:29-31 and begin memorizing verse 31.

Would you like to start flying today? Claim the promise God offers in this passage. Ask God how you can soar for him in your classes or campus.

MARCH 23   *DIRECTIONS TO DETROIT, PLEASE*

Life usually offers many options and is full of choices. For example, if you were traveling from Dallas to Detroit, there are several routes that you could take. You could travel through Little Rock and Memphis, on up to Indianapolis, and then Detroit. Or, you could take I-35 to Oklahoma City, hit I-44 to St. Louis, I-55 up to Chicago and then over to Detroit. In reality, there are dozens of different routes that you could take to Motown.

There is one road in life where no options are available. The route to God is marked clearly in John 14:6. It says, *"Jesus answered, 'I am the way and the truth and the life. No one comes to the Father except through me.'"* So many people try to find different routes to God. Maybe you are one of those who has tried to find God through Eastern religions, through meditation, through knowledge, through discovering self or nature, or through countless other dead-end routes. The one route to God leads through salvation in Jesus. He is the way to eternal life.

MARCH 24   *SANTA,*   *GET ME EVERYTHING*

Read Job 1:1-3, 9-22. Read verses 21-22 again.

Job was greater than any other man of his time. He was wealthy, successful, happy and had a beautiful family; and it was all taken away from him. Most important, he was extremely dedicated to God. When he lost everything, he continued to praise God. What a great example of the Christian spirit we should have, regardless of the circumstances.

It is so easy to feel discouraged because of a broken relationship or a problem at school or a problem at home. We often fail to recognize how truly lucky and blessed we really are. We wonder why God doesn't give us more, when we really don't deserve what we have.

Ask God to forgive you for having an expectant spirit. Thank Him for His love and the fact that you can depend on Him.

Prayer: "God, may Your desires and Your ways become my desires and my ways."

Pray this with all your heart or don't pray it at all.

MARCH 25   YOU'RE NOT AS SECURE AS YOU THINK

A man was walking through the woods on a cold January day when he happened upon a wide creek with steep banks. His mind reminisced of the days when, as a young boy, he would swing across similar creeks on vines hanging from tall trees. Ironically, he looked around and noticed a vine hanging from a tree, inviting him to relive those boyhood days. He tested the vine to see if it would hold him, backed up, and launched from the bank out over the creek. Wow! It was just like old times. He made several more swings and decided to end with one grand ride over the creek. The cold water below rushed swiftly due to recent rains. He backed up as far as he could go, ran as fast as he could run, and took flight. At the height of his swing, the vine snapped, dropping him into the icy water like a penny in a mall fountain.

So often in life, we begin to feel secure and safe, even to the point of becoming careless. We swing over danger or temptation with the attitude that we will never fall in. In I Corinthians 10:12, Paul warns us: *"So, if you think you are standing firm, be careful that you don't fall."* Don't become careless in your walk with God for the sake of a little fun or excitement.

What do you do to very strong tea? Add water, right? A process like this happens in our Christian lives at times, too. When we are first saved, that perfect combination of God's grace and our faith forms a very strong mixture. Slowly we add more and more activities to our schedule; Sunday School, choir, camp, revival, after game fellowships, and on and on the list grows.

Many times we forget to increase our time alone with God in proportion to the intensity of our involvement in activities. Too much water has weakened that once strong tea.

Never push God into the corner because you are too *busy* being a Christian. Take time to listen to what He wants to say to you through His word. Read Proverbs 5:1-2, 7. Notice what the Lord says NOT to depart from.

MARCH 27   A DOLLAR FOR YOUR SOUL

The old song of the late 1970's summed it all up in one word: "Money, money, money, money...MONEY." For some people, and for many college students and college grads, making money is a top priority in life. Someone once said that only kisses and money could be so full of germs and still be so popular! Nowadays, people will just about do anything to become rich, even if it means resorting to illegal, immoral devices.

Money is not bad. It is the obsession with money that destroys a person. The Bible makes that very clear in 1 Timothy 6:10. Consider it a warning against sure destruction. It reads, *"For the love of money is a root of all kinds of evil. Some people, eager for money, have wandered from the faith and pierced themselves with many griefs."* Can you put a dollar value on your soul? Long for God and he will prosper you with money and things money cannot buy.

Read 1 Chronicles 1:10.

You may feel your name is ordinary, but, your name can represent your reputation. How do others perceive you when they hear your name?

Fontana, California, (AP) – What's in a name? No end of harassment for a 19-year-old steel corporation employee here, who plans to enter college in the fall. His name is Richard M. Nixon. The 'M', though, is for Mark. He is a Democrat and he is not related to the former president with a similar name. "People are always saying things like, 'Hello, Mr. President, how are things at the White House?'" Richard said. "If I try to charge something at a store, I'm usually told, "Sorry, your credit's no good, ha, ha!" It isn't even funny anymore, but I try to be polite and smile once in a while."

Richard said he was stopped by a policeman one night and when he gave his name as requested, the cop snapped: "Okay, you smart punk, now put your hands on the car and don't move." After the frisking, the cop ran a check on Richard's license plate, since he'd lost his driver's licence and car registration. "When he found out I was telling the truth," said Richard M. Nixon, "he started laughing like crazy."

Nimrod means warrior. What do people associate with your name? Drug user? Alcoholic? School clown? SGA President? Pervert? Homosexual? Write down how you can begin to use your name to associate with the great things you want to accomplish.

MARCH 29   CALL IT WHAT IT IS

A great message of hope can be found in 1 John 1:9. Maybe you have heard it or read it. Take a look at what it says: *"If we confess our sins, he is faithful and just and will forgive us our sins and purify us from all unrighteousness."* This is an if/then scripture. If we do our part (confess), then God will do His part (forgive).

So often, however, we do not want to come face to face with the sin in our life and call it what it is. It's kind of like the restaurant that ran an ad that it was taking applications for a "ceramic engineer." They needed a dishwasher. If the problem we have is with lust, we must confess that to the Lord and call it what it is. If it's anger, confess it. If it's pride, face it and confess to the Lord that you have a problem with pride. God already knows. But He wants us to do our part of confessing to Him the sin we want Him to cover and wipe out of our lives.

MARCH 30   GANGS, GANGS, GANGS

Read 2 Chronicles 7:14.

Cliques and gangs have always been around, but violence has plagued the street gangs of the late 1980's and early 1990's. It is said that Los Angeles alone has 75,000 gang members.

History's greatest secret project was the secret development of the atomic bomb by the United States during the World War II. Franklin D. Roosevelt approved the project in 1939, but actual work started in 1942 under "AAA" priority. On July 16, 1945, the bomb was successfully test fired at New Mexico.

The entire project had involved over 600,000 men but for nearly 4 years the secret of the bomb was protected by their silence. Each scientist and project was assigned a code name, and "atom" or "bomb" was never mentioned in conversation.

Towards the end of World War II, the news appeared that the Germans had developed the atomic bomb. But the onrushing allied troops found the German A-bomb at its elementary stage of development.

Hitler never learned that the USA was that advanced in the development of this nuclear weapon.

The greatest secret weapon in your life is prayer. Prayer can bring forgiveness, healing to your life, and the attention of God. Are you using your secret weapon to join God's gang? Write down how you can pray and join others in prayer.

As the winter months pass and spring begins to unfold before our eyes with flowers and life, we are reminded of Easter and new life in Christ. The words in Mark 16:6 never lose their power to excite and thrill as we read, *"You are looking for Jesus the Nazarene, who was crucified. He has risen!"*

In April of 1988, *Focus On The Family* printed a story by Ida Mae Kempel in its magazine. It was about a boy named Jeremy and his teacher, Doris Miller. Although 12 years old, Jeremy was still in the second grade. He had a twisted body and a slow mind. During the spring, Doris told the story of Jesus to the 19 students, and then gave each one a large plastic egg to emphasize the idea of new life springing forth. She told the students to bring it back with something inside that showed new life. Jeremy just stared and listened. The next day, all the students were asked to show what they brought, things like butterflies and flowers. Mrs. Miller opened Jeremy's egg, only to find it empty. She was flustered because Jeremy had not understood the assignment—until Jeremy spoke up. "Jesus' tomb was empty too!" He understood new life completely.

There are consequences, good or bad, to everything we do in life. Sometimes those consequences are realized immediately, and other times they are realized later on down the road. An example of the former would be an NBA basketball game between the New York Knicks and the Phoenix Suns. A fight erupted between two players on the court, sparking an all—out brawl between both teams. Players even came off the bench to engage in the fight, one of those in his street clothes. The NBA handed down stiff fines totalling over $20,000 and suspensions for several of the players.

An example of the latter is found in Romans 6:23. It says, *"For the wages of sin is death, but the gift of God is eternal life in Christ Jesus our Lord."* The fine or the payment for our sins is death. Our death. Many people do not realize this or else have chosen to ignore the fact that we will be required to pay for our sins. That is, unless we have asked Jesus to pay our debt of sin by washing us in his blood. That is a payment that cannot be beat.

Read Isaiah 14:12-14.

Do you believe in absolute truth of a real devil? Recently in his book, *What Americans Believe,* George Barna discovered that one-third of all adults strongly agree that Satan is merely symbolic of evil, but does not exist as a true presence. Overall, three out of five adults are inclined to disbelieve in the existence of Satan. Only 25%, or one out of four, believe there is a real devil.

An old deacon who used to pray every Wednesday night at prayer meeting always concluded his prayer the same way. "And Lord, clean out all the cobwebs in my life." Another deacon had heard the conclusion so long that when the old deacon prayed again, "Lord, clean out the cobwebs," he had had enough. He jumped to his feet and shouted, "Lord, don't do it. Kill the spider. That's what needs to be done!"

Do not be confused with terminology of "symbolism" and "reality." Satan is real and we must kill his effectiveness in our lives each day through the blood of Jesus. Are you standing in God's truth or daring the devil to exist? Write down how you can kill the spiders of your life.

APRIL 3   HEARTBREAK HOTEL

A recent study showed that about a quarter of college freshmen with boyfriends or girlfriends back home break up with their high school steadies within nine weeks after starting college. One co-author of the study states, "Students who depend a lot on their boyfriends or girlfriends for their happiness are going to be most upset when they come to campus and that person is not available."

Granted, relationships can bring much happiness and the "right" person can make our liver quiver, but our source of joy and happiness as Christians cannot be linked with people, places, or things. In Romans 15:13, Paul says, *"May the God of hope fill you with all joy..."* Our happiness is not dependent on what happens around us, but what happens *within* us in our walk with God.

APRIL 4   FOLLOW ME

Read Ezra 10:1-2.

All around your campus are directions to this meeting or this event scheduled for your school. If all else fails, follow the directions.

The celebrated clipper ship Dreadnought once sailed backward for 280 miles–a technical feat unique in maritime history. The story:

In 1862 while going west from Liverpool in the Atlantic, it was struck by a strong gale which killed the carpenter, broke the rudder, and disabled the vessel. For three days, the ship wallowed in the trough of the sea.

Then the captain decided on a desperate maneuver. He furled all lead sails and all canvas on the foremast–set all square sails on the mizzenmast and threw back every sail that was set. By steering over the bow, the captain sailed the boat backward, to the nearest harbor in the Azores.

Are you following people in reverse? Are the role models leading you to safety or destroying you little by little? Ezra gave the people something to follow so that hope could come in their hearts. Are you following others or saying, "Follow me?" Write down how you can be an example to others.

APRIL 5   A METHOD TO THIS MADNESS

College students across the nation are familiar with a phrase that summarizes one month out of twelve for basketball fans everywhere: March Madness. The NCAA basketball showdown begins with sixty-four teams and ends on the "Road to the Final Four," where, eventually, one team remains as college basketball's victor. Teams like back-to-back champs Duke or the powerful Hoosiers from Indiana to obscure, Cinderella hopefuls like Coppin State fight for the National Crown. All sixty-four teams get a shot.

Paul says in 1 Corinthians 9:24, *"...in a race all the runners run, but only one gets the prize. Run in such a way as to get the prize."* Will only one person make it to heaven? Of course not. But we should live each day with the attitude that if we got one chance, one shot, we would do our best everyday to insure that we would be left standing when all was said and done.

APRIL 6   HAND ME ANOTHER BRICK

Read Nehemiah 2:17.

If someone were to drive a pick-up truck through your living room, would you leave the holes in the wall?

A curious freak of nature occurred near Learned, Kansas, November 18, 1897. The railroad station at Roasel, on the Jetmore branch of the Atchison, Topeka, and Santa Fe railroad, was swallowed up by the earth. On the morning of the 19th, there was only a dark and stagnant pool of water where, twelve hours before, there had stood a depot, a grain elevator and several small buildings. This happened during the night, when no one was near. The next day, the longest ropes obtainable were lowered into the depths of the pool without touching bottom, and sticks of timber thrown in were sucked down out of sight as by an immense undertows.

Night by night, day by day, the walls of morality and righteousness are being shattered in our nation. We, the church, have the bricks in our hands to rebuild the walls, brick by brick, person by person, class by class, day by day. We must stand in the gap. Some people talk history. Some people read about history. Very few people make history. Nehemiah made history by undertaking the task of rebuilding the walls–will you? Write down how you can build up the walls in your life and school today!

APRIL 7   LOST & FOUND

The report came in just before midnight. A group of four men had taken their cabin cruiser off the coast of Florida for a three–day fishing trip but had not returned as scheduled. The Coast Guard was notified and orders were given to begin an extensive search for the lost men. Comparable to finding a needle in a haystack, special radar equipment, helicopters, planes and boats were all used in the search. After covering 38,000 square miles of ocean, the lost vessel and men were found adrift in the Atlantic.

We must never forget that while drifting in a sea of sin, Christ launched a search for us. He came after us to rescue us, much like the shepherd spoken of in Matthew 18:12. It reads, *"What do you think? If a man owns a hundred sheep, and one of them wanders away, will he not leave the ninety-nine on the hills and go to look for the one that wandered off?"* God is willing to do whatever it takes to seek us out and rescue us when we wander from Him. What comfort we can find today in knowing He loves us that much!

APRIL 8   THE LONG ARM OF THE LAW

A popular song by country star Kenny Rogers says, "You can't outrun the long arm of the law." Read 2 Samuel 12:1-13.

Pastor Paul Tinlin of the 250 member Evangel Assembly of God Church in Schaumburg, Illinois, is getting a lot of press attention in the Midwest. It all started when Tinlin, 41, wrote a letter to a local newspaper, disagreeing with an editorial that praised the Supreme Court for in effect striking down the death penalty.

His letter stirred up sharp reaction, prompting a stiffer stance by Tinlin. "There should be swift and sure justice for those who kill—and that should be public execution, and the execution should be on prime-time TV," declared the minister. "We've got to start letting society see life for real," he explained to a Chicago reporter. "Society should know that killing isn't like on TV shows where the victim gets up and walks away when the show is over, that when real people get killed, they are dead."

Tinlin told his questioning 12-year-old daughter that seeing executions on TV "would probably make me sick, that it would be gruesome." But, said he, "murder is also gruesome, and society has to start taking it seriously." He cited a verse in Genesis: *"Whoever shed the blood of man, by man shall his blood be shed."* Maybe, he said, it's time for God's harvest law to be followed.

No one ever outsmarts God. Each of us will have to answer for our life. Are you hiding things from God? Write down what you need God to cleanse in your life.

A woman in charge of the church magazine sale fund raiser stood to make an announcement in church. "Everyone should begin their subscriptions this month so we can all expire together." We may not all die together, but the fact is death is inevitable. The Bible makes this clear in Hebrews 9:27. It says, *"Just as man is destined to die once, and after that to face judgement.."* At that point, there are no second chances.

No one knows for sure when he or she will die. Death is no respecter of age, social background, or time schedules. We must be ready to stand before the judgement seat of Christ and give an account of our life to God. Are you ready? On a tomb in a historic cathedral in Jamaica is this epitaph: "Life teaches us how to die; death teaches us how to live." Both are good teachers. Unfortunately, we are slow learners.

Living conditions at a college or university get pretty hectic. Read Genesis 8:1-9.

A hurricane is said to lift sixty million or more tons of water into the sky and can generate more power every ten seconds than all the electrical power used in the United States in one year. The hurricane that struck Bangladesh in 1970 produced a tidal wave which killed at least 500,00 people. In 1900 at Galveston, Texas, a hurricane created storm tides that swept 6,000 people to their deaths. Another 1,000 people were drowned in 1954 when a large ferryboat was sunk by a hurricane in Kakadote Bay in Japan's north island. Storms can be deadly.

Noah was in a storm, and he warned the people to prepare. However, many delayed or disagreed altogether. God spared Noah in the storm. God will sustain you as well—no matter how fierce the storm may be. The condition—stay in the ark.

Dorm life provides many storms and battles. Are you warning those in the dormitories to prepare? We must warn them of imminent danger! Write down names of those you need to warn. God will sustain them as well–if they come into the ark–the ark of Jesus.

APRIL 11   PAY DAY

You have, most probably, heard the time–worn expression "Two wrongs do not make a right." Many people utter these words, but how many actually live by them? Our society teaches that "turnabout is fair play" and that paybacks are in order. We feel it is our right to get even or to make sure that if we are hurting, the person who caused that pain is going to hurt, too.

God has not given us the right or the authority to decide how people are punished. But he has given us the responsibility of showing kindness to everyone. In 1 Thessalonians 5:15 we read, *"Make sure that nobody pays back wrong for wrong, but always try to be kind to each other and to everyone else."* When we assume the right to punish others for the hurt they have caused us, a right that is not ours, we open ourselves up to God's discipline. An eternal pay day is coming for everyone. In the meantime, we should daily strive to show kindness to everyone. It is true, two wrongs never make a situation right.

APRIL 12   "YES, LORD!"...

Once a man visited a church which was known for its spirited worship services. The man took a seat on the front row, and here's what took place. The pastor went over to the piano and, beginning to play, he said over and over, "Yes, Lord, yes, Lord." Soon a woman in the congregation joined in, "Yes, Lord!" Then more and more people joined the mighty chorus, "Yes, Lord, yes, Lord." Suddenly the congregation grew quiet and the pastor prayed this prayer: "Lord, you've heard our answer. Now tell us what it is that you want us to do!"

That kind of openness to the Lord's will is what He is looking for. Read 2 Chronicles 16:9, and write down a description of the kind of person God loves to support.

APRIL 13   THE ROAD TO HELL

Do you believe in a literal hell? Some believe we are experiencing hell now. Others believe it is merely a fictitious place. Still others believe in varying degrees of hell, or that hell will be one giant party. It used to be common for preachers to talk about hell and to preach "Hellfire and Brimstone" messages. Somehow, today, we avoid the topic and focus more on "feelgood" ideas and subjects.

However, pushing the topic aside does not change the fact that hell exists. The Bible describes hell as a place of eternal, fiery torment, and makes it clear that anyone who doesn't receive Jesus as Lord and Savior will spend eternity there. Everyone has a choice! Good morals or good intentions are not enough. Someone once said, "The road to hell is paved with good intentions." Matthew 13:49, 50 says, *"This is how it will be at the end of the age. The angels will come and separate the wicked from the righteous and throw them into the fiery furnace, where there will be weeping and gnashing of teeth."*

APRIL 14   WHAT LEADERSHIP REQUIRES

Read 1 Samuel 2:9-10.

Many people everyday are looking for something to follow. "Follow the leader" is a popular phrase, but where are they leading us? Toward or away from God?

H. Gordon Selfridge built up one of the world's largest department stores in London. He achieved success by being a leader, not a boss. Here is his own comparison of the two types of executives:

The boss drives his men; the leader coaches them.

The boss depends upon authority; the leader on good will.

The boss inspires fear; the leader inspires enthusiasm.

The boss says "I"; the leader, "we."

The boss fixes the blames for the breakdown; the leader fixes the breakdown.

The boss knows how it is done; the leader shows how.

The boss says, "Go"; the leader says "Let's go!"

There are many leaders in our world today. Music, movies, politicians, etc. What are you following? Are you a leader? God tells us that those who follow him will be exalted! Whom or what are you following today?

APRIL 15   HOW MANY TIMES?

The lyrics to a song by popular contemporary Christian artists "Whiteheart" say:

"How many times should I stand in the waves of this crashing sea? How many times must I forgive all the hurt that's been done to me? Seventy times seven."

Peter asked this same question of Jesus in Matthew 18:21, 22. *"Lord, how many times shall I forgive my brother when he sins against me? Up to seven times?"* You see, Peter was being overly gracious when he asked if he should forgive seven times. The expected, standard number of times to forgive someone according to law was three. So Peter was more than doubling the norm. Jesus' answer blew Peter and all religious law out of the water. "Jesus answered, *"I tell you, not seven times, but seventy—seven times.'"* In other words, Jesus was saying as many times as it takes, we should practice forgiveness.

Read 2 Samuel 18:33.

There are many giants in our land–drugs, alcohol, STD's, divorce, but the giant within must be conquered before the others can be slain.

The Hiroshima Bomb, though utilizing more matter due to its yet primitive process, was classified as a "nominal" 20-kiloton explosion. It obliterated but a four-square mile area, yet its blast effect was so intense that 50,000 people were killed and 55,000 more were wounded and 200,000 people were left homeless.

A subsequent report form the scientists at Kyushu Imperial University classified the effects of the bomb on the human body under three headings: (1) Instant death; (2) Symptoms like those of dysentery followed by death; (3) Throat ulcers, bleeding gums, falling hair and eventual death.

Like the bombs of WWII, there are many side effects from sin. Many view sin like parking tickets: you will get taken to jail; however, one here or there won't make any difference. The Bible views sin like cancer cells...1 or 2 do make a difference and can mean life or death. Cancer cells multiply rapidly–so can sin! What's the giant "sin" in your life that must be slain? It's a matter of life or death.

To love and to be loved is one of the basic human needs with which we are all born. And the natural desire is to settle down with one person and to share your life and love with that person for a lifetime. Ephesians 5:31 says, *"For this reason a man well leave his father and mother and be united to his wife, and the two will become one flesh."* God is very concerned and interested in this area of our lives and wants to help us avoid mistakes that could ruin our lives.

As a single college student or young adult, it is important to understand that God has a plan for you as far as marriage is concerned. We must commit the process of selecting a mate for life to God. By doing this, we can avoid some painful pitfalls. Avoid deciding too quickly or even when you are too young. Avoid the "old maid" phobia. Set high, godly standards based on God's Word and maintain those standards. Have realistic expectations. God wants the best for you. Let him help you discover who that is!

APRIL 18   JUST DO IT!

Read 1 Kings 17:15-16.

As a college student, independence is paramount. We do not like others telling us what to do. We like our own schedules and our own activities.

When a political columnist says "every thinking man," he means himself. When a candidate appeals to "every intelligent voter," he means everybody who's going to vote for him.

The widow did the unselfish thing—she gave her last meal away. It's very difficult not to be selfish, but God promises as we "just do" what he has instructed us to do, our barrels will be full also! Write down how you can be unselfish and giving to someone today.

APRIL 19   WOULD YOU FOLLOW A PERVERT?

According to the dictionary, to pervert something means to cause it to be misused, falsified, or misunderstood. *"Evidently, some people are throwing you into confusion and are trying to pervert the gospel of Christ."* Paul wrote these words to the churches in Galatia in Galatians 1:7.

Are there people today who are purposely trying to pervert the gospel of Christ, to misuse, falsify, or cause the gospel to be misunderstood? Absolutely. Many people are confused as a result. Groups like the Jehovah's Witness, Christian Science, and New Agers mix the truth of God's Word with partial truths, perverting the

Gospel and confusing believers. Some televangelists have misused the gospel for personal or financial gain, another form of perversion.

Paul rebuked the churches in Galatia for abandoning the basic, simple truth of God's Word. Even when people come along who are perverting God's Word, we have a responsibility to not allow ourselves to be thrown into confusion. How do we do this? By keeping our eyes on him, knowing His Word, and sticking to the truth of God's Word.

APRIL 20   LIFE AFTER DEATH

Read 2 Kings 4:28-37.

Death is a certain thing that each of us must face in life. Only the rapture of the church will separate some from the taste of death.

One time a poor little boy saw all the big people putting money into the offering plate. He didn't have anything to give but five marbles in his pocket. He put them into the plate! People around must have smiled as they saw the strange offering drop in. After the meeting, one of the deacons asked the child if he wanted his marbles back. "Oh, no I gave them to the Lord Jesus."

As the story was told from one to another, a rich man said, "I'll give a hundred dollars for that boy's marbles." That was another kind of miracle change, from 5 cents to $100.

You may feel you have not much to give in college life, but if you know Christ and the power of the Holy Spirit, you have much to give—the breath of life, just as Elisha did. A little is much with God. Who is dying around you, in need of the breath of life that you can give?

APRIL 21   GOD'S LOVE PREVAILS WHEN WE FAIL

Have you ever been betrayed by a close friend? If so, you know the deep pain, loneliness, and hurt that can result. Jesus surely felt these same feelings when betrayed by his close friend Peter. Imagine the incredible guilt Peter must have felt. He illustrated the ultimate in failure. Here was a man who was chosen as a disciple, walked on water to meet Jesus, performed great miracles, was called "petros" or rock by Christ, prayed with Jesus in the garden, and even defended Jesus in the garden. Then, only a few hours later, he denied even knowing Christ, not once but three times.

Failure is not easy to deal with, especially when we fail our Heavenly Father. In Mark 16:7, two words demand our attention. *"But go, tell his disciples and Peter..."* After his miserable failure, God wanted to be sure that Peter knew he had not forgotten hem, and that he still knew Peter's name. Do you feel you have let God down? Be encouraged. He still knows your name, and He has not forgotten you!

APRIL 22   IT'S NOT THE DEVIL'S FAULT

Read 2 Chronicles 7:14.

Many people across our land blame the devil for the economic and social troubles we have. However, according to 2 Chronicles, it's the church that is to blame for not doing its part.

In the early 1960's, a wave of feeling known a "Black Pride" swept across America. Everything about the black culture, from hair texture to history, had taken a back seat to that of the white majority. Over time, a dramatic rediscovery of black heritage occurred. Textbook publishers issued new editions that for the first time included the stories of black Americans: a soldier who rowed George Washington across the Delaware, a scientist who perfected the art of blood transfusion and educators who founded black colleges. Hope was restored to their race.

If America has hope, it's not in Washington D.C. or Wall Street. Hope lies in those who profess Jesus as Savior and the prayers that they pray!

Are you praying for our nation? Write down how you can pray in hope for America.

A conversation was overheard between two college students. "I was so embarrassed last week when my parents came to visit," said Craig. "I took my dad to the campus arcade, and he tried to play the change machine!" Have you ever been embarrassed by something your dad did? Maybe the time he yelled at the umpire during your Little League game, or questioned your date for an hour before letting you go. We shrink back in shame.

Sadly enough, many Christians are embarrassed of their heavenly Father for no reason. Christ never ridicules us or gives us reason to feel ashamed. That type of ridicule and shame comes from a world that has rejected Christ. Jesus said in Luke 9:26, *"If anyone is ashamed of me and my words, the Son of Man will be ashamed of him when he comes in his glory..."* One of the greatest honors in all the world is to be called a Christian, a child of God. Hold your head high, your shoulders back, and let the world know you are proud to be His!

Read 1 Chronicles 29:26-28.

King David's life was one of ups and downs. From shepherd boy to shouting king, to death and the passing of the mantle to Solomon, his son. From a pimple–faced boy, to a pitiful man, David made the history book.

The story is told of Gordon Maxwell, missionary to India, that when he asked a Hindu scholar to teach him the language, the Hindu replied: "No Sahib, I will not teach you my language. You would make me a Christian."

Gordon Maxwell replied, "You misunderstand me. I am simply asking you to teach me your language."

Again the Hindu responded, "No, Sahib, I will not teach you. No man can live with you and not become a Christian."

What a heritage David left! What will be known of your life when it is finished? When people think of you, do they think they will become Christians by associating with you? Some talk about history, some read about history, but very few make history. What will the history of your life be? Write down how you can influence someone today!

APRIL 25   STORM SHELTERS

Years ago in Edmonton, Alberta, kids were given an address when they arrived in grade school of a house that was on their way home. Winters were harsh and the blizzards sometimes caused the temperature to fall to 30 or 40 below. The address was the location of a "Storm Home." If it was too cold to walk all the way home or if you felt you could not go on, you could stop in and get warmed up with cookies and milk.

God has given us each the address for his "Storm Home." The address is Matthew 11:28. It says, *"Come to me, all of you who are weary and burdened, and I will give you rest."* Our place of shelter and rest is in the strong arms of the Father. When the storm rages, or we feel we can't go on, we simply stop in God's Storm Home and find rest and safety.

APRIL 26   SING, SING A SONG

Read Psalm 150.

Have you ever sung in the shower? It has been said most people do because of the wonderful resonance of the bathroom walls. Singing is a wonderful expression of internalized emotions.

On the way back from Mount Higashi, where father and son had gone to enjoy themselves, the father got into a very good temper and entered the village singing. He went past his own house, and the son said, "But Father, there is our house!" The father looked quite unruffled and said, "Yes, but if I go in now the song will be in the middle."

What song are you singing today–the world's or those which come from above? Ask God how you can sing for Him today, and sing a song for Jesus!

APRIL 27   GRACE THAT IS GREATER

Dave Dravecky played in the backyard of his home in Boardman, Ohio with his dad and dreamed of one day pitching in the major leagues. That dream became a reality, but in October of 1988, while pitching for the San Francisco Giants, Dravecky underwent surgery to remove a cancerous tumor from his pitching arm. Doctors said he would be lucky to play catch again in his backyard, but on August 10, 1989, Dravecky did pitch again for the Giants and won.

In June of 1991, after the tumor returned, after more surgery, after more treatments, Dave Dravecky's pitching arm was amputated. Is he bitter, angry or mad at God? Listen to Dravecky's own words, written in his book *When You Can't Come Back*: "I'm getting through it because I have a Father in heaven who is a great giver. He is where I find the grace...I don't earn it. I don't deserve it, I don't bring it about. It's a gift. And that is how I am able to cope with the 'tragic irony' of losing my arm." 1 Corinthians 15:10 says, *"But by the grace of God I am what I am..."* His grace is truly amazing!

APRIL 28   THE FOUNTAIN OF YOUTH

Read Psalm 103:1-5.

The pursuit of youthfulness and strength has been an eternal quest from centuries past. In St. Augustine, Florida there have been reports that the fountain of youth has been discovered. One drink of this fountains cool, flowing waters will invest in you the gratuities of long, strong, healthy life–NOT!

Unfortunately, many have tried and failed at attaining immortal youthfulness. Psalms says to be renewed in our strength is to

"Praise the Lord."

Worship and praise renews our spirit with God and gives us "youthful, new" strength to stand for God in an ungodly world. Ask God how you can praise and worship Him so your youthful appeal and disposition can stay with you a lifetime!

APRIL 29   DO YOU WANT YOUR BLESSINGS BUTTERED?

Okay, all you popcorn–inhaling college students, listen up. You are a part of a much larger group in the United States whose love for popcorn consumes 700 million pounds annually. Sound like a lot? Let's say that all that popcorn was poured on top of Rochester, Minnesota. This 23-square-mile city would be buried under 1800 feet of popcorn.

Now, imagine that you are Rochester, and the popcorn represents God's blessings. God's desire is to pour blessings upon his children. Ephesians 3:20 says that God is *"...able to do immeasurable more than all we ask or imagine..."* Trust God today for his abundant blessings to be poured out on you.

APRIL 30   THEY ALL LOOK ALIKE

"If you've seen one, you've seen them all." This is a popular phrase. The question is, are you like everyone else around you?

To what are you presently conforming? Who or what do you plan to be like 10 years from now? How concerned do you think God is with whom you are developing close friendships ?

Read 2 Chronicles 20:35-37.

All of the ships that Jehoshaphat had built were destroyed by God for one reason. He was involved in a relationship that God did not approve of. Your efforts may also be useless if you are not involved in the right relationships today.

Ask God if he approves of your relationships, and be obedient to his instructions.

Read Proverbs 3:5-6.

Have you ever had someone say, "You can trust me," only to have them let you down?

The king of Italy and the king of Bohemia promised John Huss safe transport and safe custody. They broke their promises, however, and Huss was martyred. Thomas Wentworth carried a document signed by King Charles I which read, "Upon the word of a king you shall not suffer in life, honor, or fortune." Shortly afterwards, however, his death warrant was signed by the same monarch. "Put not your trust in princes," were his last words.

"It is better to trust in the Lord" than in anyone or anything else.

Many times people will let you down. Disappointment sets in and we lose confidence. However, God will never let us down. Write down how you can trust God today, and how others can trust you to keep your word and actions true.

Read Nehemiah 2:17.

There are all types of construction around us today. Colonial, Victorian, traditional, Grecian, Gothic, etc. There's something wonderful about building a new building.

Woolworth conceived the idea of the Five and Ten Cent store. That was different. His fortune was measured by millions when he passed away. Wanamaker conceived the idea of one-price to everybody in his retail stores. That was different, for at the time, he put this policy into effect it was directly contrary to accepted practice throughout the country.

Ford determined to build a light, cheap car for the millions. That was different. His reward came in the greatest automobile output in the world. Human progress has often depended on the courage of a man who dared to be different.

Men and women have built empires, but Nehemiah built a wall, a structure that led God's people to victory and success in God again. The wall was destroyed, but brick by brick, Nehemiah put it together again. Are the walls of your life shattered or strong? Ask God how you can rebuild your walls so you to can be in victory throughout this semester. Hand me another brick!

MAY 3   MINUTES 'TIL MIDNIGHT

It is amazing to think that people who were born before 1945 were born before television, penicillin, polio shots, frozen foods, Xerox, plastic, contact lenses, frisbees, and the Pill. These folks were before radar, credit cards, split atoms, laser beams, and ballpoint pens. They were before pantyhose, dishwashers, clothes dryers, electric blankets, air conditioners, drip dry clothes, and before man walked on the moon.

Times have certainly changed, and at a whirlwind pace. How practical the Bible is when it says to us in Romans 13:11, *"...do this, understanding the present time. The hour has come for you to wake up from your slumber..."* God help us to wake up and realize that our nation has seen more than just technological change. Our nation has witnessed rapid, drastic moral and spiritual change. We must understand the time and allow God to stir us from sleep, determined to do our part to bring about change through salvation in Christ.

MAY 4   FINDERS WEEPERS, LOSERS KEEPERS

Read Ezra 10:1.

It seems very easy to point out the faults in others while ignoring our own. It's called the "log eye syndrome" where we worry about the speck in someone else's eye when we have a telephone pole in our own.

When Chaplain McCabe, who later on became a bishop, had set out to raise a million dollars for missions he met many disappointments and was often greatly discouraged. One day while going through a mail that was particularly discouraging, he finally came across a letter from a boy, from which fell a badly battered five-cent piece. The letter, in a boyish scrawl, and liberally punctuated with blots, read:

"Dear Chaplain McCabe:

I'm sure you're going to get a million dollars for missions. And I'm going to help you get it too. So here's a nickel toward it. It's all I've got right now, but if you need any more, you just call on me."

This became one of the Chaplain's most effective stories in his money-raising campaign, and by it he was eventually able to reach his goal. A boy's nickel, multiplied, became a million dollars. He helped far more than he knew.

This little boy internalized a need. We often say, "Let someone else do it." Ezra understood. He had to apply it to himself before others could understand.

Today, look not at others on your campus, but at your own life and ask, "What do I need to change or give away?" Jesus said if you lose your own life, you find it —finders weepers, losers keepers!

MAY 5 THE STING OF DEATH

A young first grader teased one of his female classmates with a half-dead bee and then dropped the bee down the back of her shirt. The last thing he saw before the principal escorted him out of class was the girl running down the hallway, screaming in tears. Fortunately, most bees only sting once, leaving their stinger in their victim.

Jesus took the fiery stinger of death once and for all when He died on the cross. In 1 Corinthians 15:55, we read, *"Where, O death, is your victory? Where, O death, is your sting?"* Death is still present, and we will all die, but for the Christian, the pain and poison of death is gone. The sting of death is removed. Death has no victory over the believer, but we have victory over death.

Read Esther 4:15-16.

What makes a hero? Ask one and you may find surprising nonchalance. "I'm sure others would have done the same thing," they say or "I was just in the right place at the right time." People become heroes because they take quick action when others stand and watch.

People who supervise others have decisions to make every day. They range from minor incidents to vital problems–whether to reprimand someone for being five minutes late or to setting quotas on a production line. Each decision saps energy and requires effort. But they must be made. Otherwise, management is failing to manage.

If an important policy or big sums are involved, take time to think, but not to tremble. Consider all of the angles. Then, if it's up to you, make the best decision you can and let the chips fall where they may. Your decisions will never be 100% correct; neither will those of your subordinates. If you want action, not indecision, make it clear that you don't expect perfection every time.

Good leaders want their managers to manage. They want action and a winning average–not perfection. They want an organization that moves instead of one that rests on it laurels from fear of making mistakes.

Esther was the type to take action. Are you involved in ministry on your campus? Do you actively participate or complain at how liberal the administration is? Take action today! Heroes act! Ask God how you can be involved in leading your campus in ministry. Act now!

MAY 7   *WITHSTANDING THE TEST OF TIME*

No other book in all of history has withstood the tests of time, scrutiny, and research as the Bible has. Many have referred to it as the "Owner's Manual" for the Christian. It is still as relevant and powerful today as it was hundreds of years ago. Why is that so? The answer is found in one scripture, 2 Timothy 3:16, which says, *"All scripture is God-breathed and is useful for teaching, rebuking, correcting, and training in righteousness."*

This verse gives the origin of the term inspiration, which literally means "God-breathed." God inspired the writers, or breathed into them the inspiration to write the sixty-six books of the Bible. This is why God's Word is useful to teach, rebuke, correct, and train us. This "Owner's Manual" contains everything we need to make it as winners in this life.

MAY 8   "I DO"

Read Proverbs 21:9.

The two most important decisions in life are (1) to place Jesus as the Lord of your life and (2) the person to whom you will marry. The wrong decision in either of these will mean possible frustration and defeat all of your life.

The bride and bridegroom were handcuffed to each other during the ceremony. As they stood there before the minister, one arm of the bridegroom was handcuffed to the adjoining arm of the bride.

It happened at Cambridge, Massachusetts, where Fernan Lowe was tipped off that some friends of his were up to a practical joke. They planned to abduct the bride at the wedding. But Fernan fooled them. He fixed it so they'd have to abduct him also. That's why they were handcuffed at the altar. They went off on their honeymoon still handcuffed together.

By choosing the wrong mate for life (men choose women and women choose men by God's design) you can be handcuffed to a life of disaster and hurt! Whom are you thinking of marrying? Think it through–God wants you happy in marriage–handcuffed to his joy–no anger or frustration.

Take a look at James 1:19-20. This scripture provides the solution for one of our nation's most severe problems: violence. In 1991, more than 11,000 Americans were killed in homicides committed by high-school-age attackers. The number of violent acts viewed during prime-time television has increased at a mind-boggling rate. The results? Drive-by shootings, gang wars, theft, metal detectors in our public schools, riots in our streets, and the list goes on.

The solution? James says in these scriptures that *"Everyone should be quick to listen, slow to speak and slow to become angry for man's anger does not bring about the righteous life that God desires."* We have it backwards. We are quick to speak and become violent, and slow to hear. God help us to heed His words and bring an end to the growing problems we now face. The solution begins with you. Let's turn this thing around.

Read Proverbs 1:22-33.

Have you ever had your parents or someone say, "Listen to me or else"?

Telephoning TV viewers after a newscast, Andrew Stern, a former ABC News staffer now on the journalism faculty of the University of California at Berkeley, found that 51% of those who had listened could not recall even one of the show's 19 items. Among all those called, the average memory rate was one item. The calls were made over a period ranging from immediately after the show's sign-off to 3-4 hours later. Not surprisingly, the lead story was the most remembered.

Far and away the most quickly forgotten material was the show-ending commentaries. Stern blames the poor retention rate on "disrupting factors," especially dinner. His recommendation: the networks should shift their major newscasts of the day to 10:30 PM.

The world around us is screaming for us to listen to their liberality and humanism. God is shouting equally loud to listen or else. Proverbs warns that not listening to God's Word or His still small voice will have devastating impact.

Write down how you can and will listen to God's voice today.

MAY 11   "I DON'T GET MAD...I GET EVEN"

That saying has covered the front of many T-shirts and has been seen on plaques and bumper stickers all over America. If someone cuts in front of you in traffic, lay on the horn and yell a few obscenities. If you don't get invited to the big social, spread a few lies about those who are hosting. If your roommate borrows your clothes without asking, take enough of her things to justify the "trade-off." Just think, if only Jesus would have come down off the cross and cleaned house, or retaliated when mocked by Pilate.

Instead, Jesus said in Luke 6:28-29, *"...bless those who curse you, pray for those who mistreat you. If someone strikes you on one cheek, turn to him the other also."* Jesus knows that revenge looks sweet, but ends up damaging us in the end. He knows that more is accomplished by being self-controlled than by being out-of-control. It does feel good to get even, but we must follow Christ's example and go the second mile to forgive and forget.

MAY 12   BUDGET DELUXE MOTEL

Read Psalm 1:1-5.

Everyone is looking for a special savings on purchases today, 10%, 30%, 50% off. Not all that glitters is gold, and cheap deals don't always guarantee quality. Happiness is not always inthe discount.

Where is happiness found? John B. Rockefeller, a Christian

millionaire, said, "I have made many millions, but they have brought me no happiness. I would barter them all for the days I sat on an office stool in Cleveland and counted myself rich on three dollars a week." Broken in health, he employed an armed guard.

W.H. Vanderbilt said, "The care of 200 million dollars is too great a load for any brain or back to bear. It is enough to kill anyone. There is no pleasure in it."

John Jacob Astor left five million, but had been martyr to dysplasia and melancholy. He said, "I am the most miserable man on earth."

Henry Ford, the automobile king, said, "Work is the only pleasure. It is only work that keeps me alive and makes life worth living. I was happier when doing a mechanic's job."

Andrew Carnegie, the multi-millionaire, said, "Millionaires seldom smile."

Psalms tells us that true happiness means being grounded in the Lord Jesus Christ. Are you shopping for spiritual discounts, or are you going to the Rock for stability? Ask God how you can be firmly grounded in Him.

MAY 13   AND THE ANSWER IS...

Put the coffee on. Break out the donuts. It is finals week, and the process of preparing for countless exams begins. It is an awesome challenge to ready yourself to recall pages of notes and information in order to answer correctly the question on the test. The reward of completing the test and the course makes it all worthwhile.

As followers of Christ, we hold the answers to some of life's most difficult tests and trials. In 1 Peter 3:15, we are challenged to *"Always be prepared to give an answer to everyone who asks you to give the reason for the hope that you have."* We have to take the steps of preparation in order to answer correctly when someone inquires about our hope in Christ. The world will see a difference in those who are living for Jesus and will want to know what that difference is. Be ready to tell them!

Read Proverbs 24:33.

Do you know people you would call lazy? On a college campus, laziness can lead to academic probation or worse, being expelled from school all together. Laziness can rob you of honors that should have been yours.

Dr. Margaret Mead, distinguished anthropologist and author, made a very interesting observation in an address not long ago. She pointed out that for a long time it was the universal custom to say on parting: "Good-bye," which is a shortened form of "God be with you." Today it is quite common instead to say: "Take it easy."

The Proverbs tell us that laziness will steal from us like a bandit. One day, the lazy will look around and ask, "Why do I not have what others do?" There are two ways to climb an oak tree. You can climb it or you can sit on an acorn. The choice is yours. Write down how you can overcome laziness on your campus and in your walk with God.

Cars are a status symbol. This is especially true among youth and college students. When someone pulls up on campus in a bright red BMW or a jet black Dodge Stealth, they attract immediate attention. A cheaper car would carry them wherever they needed to go just like a more expensive one, but the status would be lost. In Acts 2, God blesses a room full of people with an incredible gift that is still available today. It says in verse 4, *"All of them were filled with the Holy Spirit and began to speak in tongues as the Spirit enabled them."* This baptism in the Holy Spirit was not given so we could wow others with miracles and great power. God baptizes people today in the Holy Spirit to empower them as witnesses. We can never use the power God has given us as a status symbol, but simply to bring others to salvation. It is a gift God wants us to have. Ask Him for it, so that you can be more effective than ever before as a disciple of Christ.

Read Proverbs 14:12.

There is popular phrase today–just do it!" This philosophy can lead to a path of uncharted territory if we blindly follow what we just feel like doing.

Announcement over the loudspeaker at Washington's National Airport: "Attention, please. Will all Piedmont passengers who have not done so please do so immediately?"

Can you imagine the confusion? What do I need to do? If we do not allow the Holy Spirit of God to quietly, day by day, tell us what to do, be assured those enemies of God will assert all their power to direct us in the wrong activities.

What are you doing? Will it please God or those around you? Write down what you can "just do" to please God in your classes.

People used to say of Paul "Bear" Bryant, the legendary coach of the University of Alabama, that he could walk on water. No one ever actually saw him do it, but we do have an incredible account of a man who did. Matthew 14:29-30 tells the story of when Peter walked on the water to meet Jesus. It says, *"Then Peter got down out of the boat, walked on the water and came toward Jesus. But when he saw the wind, he was afraid and, beginning to sink, cried out, 'Lord save me!'"*

We often hear people pointing out Peter's mistake of taking his eyes off Jesus and looking at the waves around him, but rarely do you hear Peter commended for having enough faith to get out of the boat in the first place. No one else jumped out. Jesus wants us to seekHim so intently that we will be willing to take risks, to step out in faith to meet Him. Get out of the boat today!

Read Isaiah 5:7.

Just tune into the TV any day for the 6 o'clock news and you will see we live in a violent world: murder, theft, rape, assault and the beat goes on. In Los Angeles alone, there are 2 murders every day of the year.

When the San Francisco Giants lost a baseball game, Gerald Bishop lost his temper and pumped 17 rifle bullets into his television set. The bullets went through the walls of his mobile home and penetrated the walls of a house 300 yards down the road. After his arrest, Bishop asked policeman John Grimes: "Haven't you ever wanted to shoot your TV set?"

Isaiah could well be writing about the 1990's. Violence abounds around many universities. God is still looking for righteousness. Write down how you can stand for righteousness in your school or community. Righteousness exalteth a nation.

"Start jogging!" Coach Bell would stick his head out of his office and yell this command to the basketball team at the start of every practice. Definitely not a pleasurable experience for the players, but as Hebrews 12:11 states, *"No discipline seems pleasant at the time, but painful. Later on, however, it produces a harvest of righteousness and peace for those who have been trained by it."*

Those endless laps around the gym caused literal pain for the players. The training was grueling, and at times, it seemed useless. When the season began though, the team had an edge over the other teams because they were in such good shape. The discipline and training God requires of us is often painful and grueling. At times, it may even seem useless, but we have a guarantee from God's Word that it will produce righteousness and peace in our lives.

Read Job 6:15-21.

You've heard the nursery rhyme, "Sticks and stones will break my bones but words will never hurt me." We have found this to be false. Words can cut like a knife deep to our innermost fears. We must guard what we say. Being politically correct sometimes is more important than what it does to the self worth of someone being spoken to.

One kiloton (KT) is the equivalent of 1,000 tons of TNT. One megaton (MT) is the equivalent of 1,000,000 tons of TNT. The first atomic bomb exploded in New Mexico was about 19 KT. The bomb of Hiroshima was a 20 KT bomb. The largest hydrogen bomb that has been exploded was a 100 MT bomb by the Russians in the 1960s. The force of that bomb was so powerful that measured pressure pulses from this bomb went around the world two times. Though the 100 MT bomb was 5,000 times as powerful as the one on Hiroshima, scientists are discussing bombs of 10,000, 20,000, or even 1,000,000 MT.

How powerful is a 100 MT bomb? All the gunpowder , TNT, dynamite, and nitroglycerin made sine the discovery of gunpowder are not equal to 100 MT. To equal a 100 MT bomb, you would have to drop a 20 KT bomb every day of the year for 13 years. All the explosives used on both sides in WWII were less than 3 MT.

Words can be explosive and devastating. Job's friends did not help him. They hurt him very deeply. Ask God how you can speak life and encouragement to someone, even an enemy.

MAY 21   BURNT OFFERINGS

Jim and Cathy were fresh out of college and only one month into marriage when Jim's friend, Ryan, asked, "How is the married life?"

"Great!" Jim replied. "Cathy treats me like a god!"

"Wow, what does she do?" Asked Ryan. "Well," said Jim, "she prepares a burnt offering every night."

In the old Testament, the high priest would prepare a burnt offering and offer it to God as worship, as a sweet fragrance. Now, because Jesus became the final sacrifice, we do not have to go through a high priest to worship God. We can worship Him directly, no matter what age, race, sex, or social status a person may be.

Hebrews 13:15 tells us to *"continually offer to God a sacrifice of praise–the fruit of lips that confess his name."* Our praise is offered to God as a sacrifice, as a sweet fragrance. This scripture speaks of praise that is spoken from our lips as a continuous flow of worship to the Father.

MAY 22   TOUCHDOWN OR PENALTY

Read Ecclesiastes 11:9.

On any given Saturday in the fall of the year, college football is in the spotlight. ESPN, ABC, NBC and CBS all cover the biggest games of the week! Everyone loves a touchdown but is disgusted with the penalties levied against their team.

Twain was a distinguished looking figure in his later years. One day he was strolling in the park when a little girl pattered up to him and asked if she could walk with him. Highly flattered, Twain told her stories for an hour, then gave her a nickel and said, "Now run along home, and when you grow up, you can tell your friends you once walked with Mark Twain." "Mark Twain!" Echoed the little girl, bursting into tears. "I thought you were Buffalo Bill!"

Thinking she had scored, she was penalized. Many college students, today, have as their motto, "'Do as thou wilt' should be the whole of the law." In other words, Go for it! While many think their life is a touchdown, before God they will be penalized. Are you scoring for God ? Write down how you can do God's will on your campus. You can always score with God on your team!

For centuries, men and women alike have been trying to understand and define love. One of the greatest thrills in life is falling in love with that special someone and pledging your love to him or her for a lifetime. Contrary to the popular song in the late 80s, "What's Love Got To Do With It?," love is more than a second–hand emotion.

1 Corinthians 13 is often referred to as the Love Chapter and verses 4-7 give us a clear understanding that love is a choice more than it is a feeling. *"Love is patient, love is kind. It does not envy, it does not boast, it is not proud. It is not rude, it is not self-seeking, it is not easily angered, it keeps no record of wrongs. Love does not delight in evil but rejoices with the truth. It always protects, always trusts, always hopes, always perseveres."* To do these things, actions that define love, one must make a willful choice. Love never fails!

Read Song of Solomon 8:7.

Many people refer to the cruise ship, "The Love Boat," as the ultimate vacation for lovers. There are 1-900 love lines and dating services all to match you with a compatible mate to experience the exhilaration of being in love.

One of the most striking and expressive pictures of Rossetti is that entitled "Found." The story behind the picture is a touching one. A country boy and girl fell deeply in love. They pledged to each other their deathless love.

The girl gives in to evil influences and is lured to the big city. There she sinks into a sewer of sin, throwing over her pure and happy past and even trying to forget the one she loved. He, however, remains true, seeking her everywhere. One day on Blackfriars Bridge, he meets a gaudily dressed woman. He seizes her by the wrist and tells her of his continued love. The one he loves has been found.

Song of Solomon is an erotic story of erotic lovers and their love for each other. Love is not a feeling; it's a decision! To decide to love, regardless of looks, position, wealth or knowledge, but to love because I decide to love you. "The Love Boat" is a 60 minute TV show. Love was meant for a life time!

Write down how you can love your parents or friends or spouse today!

MAY 25   A DISAPPOINTING FINALE

The words of Jesus in Matthew 6:19-20 ooze with such wisdom. He says, *"Do not store up for yourselves treasures on earth, where moth and rust destroy, and where thieves break in and steal. But store up for yourselves treasures in heaven..."* Definitely not the message of today, where greed and materialism are common. So many people will be empty and disappointed when they learn their riches and wealth will mean nothing in the end.

A startling, funny thing happened in April 1986 that illustrates this. On a prime-time, national television program, Geraldo Rivera planned to unseal Al Capone's secret vault beneath Chicago's Lexington Hotel to reveal money, diamonds and other riches. Experts blasted through the walls and the vault was finally opened. All it contained was two empty gin bottles. Invest in God's kingdom. The returns are unbeatable.

MAY 26   PARTY NAKED

Read Isaiah 6:1-8.

Parties are a big part of fraternities and sororities in college. A popular phrase is "Party naked." Some toga parties are just that–a naked party.

In his book, *I'm OK–You're OK*, Dr. Thomas A. Harris says that there are four basic attitudes toward life. The newborn infant feels helpless in relation to his all-powerful parents, dispensers of food warmth and happiness, and soon gets the idea, "I'm not OK–you're OK." He either retains this outlook or acquires another.

When he realizes his parents' "fallibility" he may think either, "I'm not OK–you're not OK" (an attitude of hopelessness–the individual is worthless and so is everyone else), or "I'm OK–You're not OK" (a psychotically criminal conclusion–the attitude that only the individual is right).

The mature adult then grows into the attitude that both he and others are basically worthwhile. "I'm OK–you're OK."

Life is more than a party. Isaiah realized in one fearful moment standing before God–I'm naked–but it was no party. He felt unprepared. Are you prepared to stand before God? We all will very soon! Ask Jesus to come into your heart in a fresh way so that you can go dressed and ready to the big party–Heaven's awesome celebration!

MAY 27   STRESSED OUT!

Have your mid-term stress levels risen to the danger zone? Are finals causing you to wonder if these are your final days? Then read Phillippians 4:6. God did not intend for our bodies and minds to be "stress factories" or nuclear stress plants. Some 80% of all illnesses are somehow related to stress. Sure, we are all faced with enormous pressures and responsibilities at times, but we must learn to take a daily dose of the remedy God gives us for handling anxiety.

The scripture prescription reads: *"Do not be anxious about anything, but in everything, by prayer and petition, with thanksgiving, present your request to God."* Do your best, and let God do the rest. In every situation, every trial, talk to God, and thank Him in advance for answers and provisions.

Read Ecclesiastes 3:17 and Acts 5:1-11.

They had been married 25 years. Then our of the blue, she wanted a divorce. He couldn't believe it. Why, after all these years? She admitted she was leaving him for another man, the Realtor who had helped them find their new retirement home. Since they were now "in love," there was no turning back—they had to be together. On the day she was to leave, her husband spoke for the last time, telling her good-bye, and then broke into sobs. She felt uneasy and hurried to get her things together. She left for her new "love's" home. Two weeks later, her "new love" was seized with a heart attack, and the fling was abruptly ended. He lingered for a few hours, and then he died. Was it worth throwing away 25 years? Judgement had come.

Write down what you need to change to avoid God's judgement.

How would you like to be one of two people crammed into an airplane cockpit the size of a phone booth without escape for nine straight days? (Remind you of dorm life?) Well, that is exactly what Jeana Yeager and Dick Rutan did on December 14, 1986. They reached a great aviation milestone by making the first nonstop circumnavigation of the globe without refueling in an odd-shaped experimental aircraft called the "Voyager". For nine days, three minutes, and forty-four seconds, these two pioneers endured claustrophobia, fear, storms, sickness, and fatigue.

What are some things that you must endure today as a Christian young person? Fear? Fatigue? Storms? Hardships are certain and 2 Timothy 2:3 says, *"Endure hardships with us like a good soldier of Christ Jesus."* Notice the words "with us". We are in this thing together, and we will one day land safely in a new land, heaven.

"God has a perfect will for your Life." How many times have you heard that before? How many times have you heard someone say, "Here is how you can find God's will for your life"?

One sure way to know God's will is to stay in His Word and learn about Him. Knowing God is the first step to knowing his will.

Read Psalm 32:8. Allow these words to penetrate your heart. Be patient and remember God has a perfect plan for your life. Allow Him to bring it into your life one day at a time. Ask God to show you His will for you today.

When James Gordon Bennett sent Henry M. Stanley to search for David Livingstone in Africa, he said, "Draw on me for a thousand pounds today to provide your equipment, and when that is exhausted , draw on me for another thousand, and when that is done, draw another; but find Livingstone."

God authorizes us to draw on Him. When one day's supply is exhausted, we are to draw another and then another and then another. Ask God His will in reaching your campus today.

The fruit of a man's work is most often something visible, like a product or material. The fruit of a man and woman's love for each other is often a child. Likewise, our very lives produce visible attitudes and actions. That does not mean we are all fruitcakes. Not all fruit is good though. Some fruit is rotten or spoiled, giving us just cause to conduct a fruit inspection of our lives.

We should examine regularly the kind of fruit our lives are producing. Galatians 5:22 says, *"the fruit of the Spirit is love, joy, peace, patience, kindness, goodness, faithfulness, gentleness and self-control."* Notice it does not say the "fruits of the Spirit." It says the "fruit," singular, meaning this is a package deal. When a person walks in fellowship and communion with God, this fruit, including all of these characteristics named, will be the visible life-style.

In late 1992, a vicious hurricane named Andrew slammed into South Florida and Louisiana, leaving thousands of people without homes. Its intensity was fierce and destructive, much like the storms of life that sometimes threaten to destroy us and leave us alone and afraid.

Do you have storm insurance? No, not through Allstate or State Farm, but insurance that provides protection from life's storms? The lives of young people, young couples, college students and young adults are often characterized by stormy days. The disciples, young and sincere, could relate. They found themselves in the middle of a literal storm in Mark 4:35-41, afraid and unsure of the future. In verse 39, Jesus *"got up, rebuked the wind and said to the waves, 'Quiet, be still!' Then the wind died down and it was completely calm."* Jesus provides great storm insurance. Let Him speak to the wind and waves that surround you and restore peace to your life.

JUNE 2   LOVE IS SPELLED G.I.V.E.

How often do you remember a new person's name when you meet him/her? Do you usually think about other people, or do you think more about yourself? Do you really know how to love?

Read 1 Samuel 18:1-5.

Jonathan had a tremendous love for his new friend. Philippians 2:3-4 states, *"Don't do anything from selfish ambition or desire to boast, but be humble toward one another, always considering other better interest, and not just your own."*

When you are around people, are they thinking about you? They are probably thinking mostly of themselves. That is the world's way. God would much rather you be conscious of and sensitive toward others.

God wants to use you to meet others' needs. He wants you to learn to love as Jonathan loved David– *"as much as he loved himself."*

Jonathan proved his love by his actions. He even gave David his clothes and armor.

Are you willing to prove your love for a friend through action? Write down how you can show others at school how you love them!

JUNE 3   NOW THAT'S A BURGER BASH

What are you hungry for? Anything but school food, right? Thank God for fast food! A study once showed that Wendy's was the fastest of fast food, delivering a hamburger, fries, and drink to the drive-through in 46 seconds. The fast food capital of the world is Hong Kong, where the world's largest Pizza Hut served 4000 customers in one day. The city's 35 McDonald's serve 70,000 customers each day.

How's your spiritual appetite? Many young people today are not making it in their walk with God because their appetite is for the wrong things. Jesus said in Matthew 5:6, *"Blessed are those who hunger and thirst for righteousness, for they will be filled."* If we have an appetite for Godly things, we will be satisfied, but a fleshly appetite is insatiable. Let's fill ourselves up with righteousness. It is "All U Can Eat."

JUNE 4   JUST START THE CAR

A car with a full tank of gas will go no where until a simple task of starting the engine is performed!

Do you think you must really be some great person for God to use you in a great way? Samuel, who was the last judge, was instructed by God to choose Saul to be the first king of Israel. He belonged to the smallest tribe and the least important family, yet God chose him.

Read 1 Samuel 9:15-21 and 10:1.

Saul became a great and powerful leader used by God. God uses people of all types. He has chosen each of us to be His children and wants us to spread our Christian love and joy to others.

Some of the people I see used today by God are the last people I would have chosen. God knows who has a willing heart. He knows who wants to be used. If you are willing, I guarantee you, he'll wear you out!

Are you willing to let God use you? Write down one unique way you think God might use you for His glory in the near future. Think big, and God will use you in a big way.

You will never get down the road until you start the car. Ask God where you can start today to do what He has already shown you.

JUNE 5   *MOVING ON UP*

Do you remember what it was like to finally move to the upperclassmen's dorm? Or to move to a nicer house or apartment as a kid? That is exciting. One day soon, we will all move to a new house that would blow the mind of any real estate agent.

Many different views and opinions exist about heaven. The Bible makes it clear that Jesus has gone to prepare a place for the righteous, and that one day He will return to escort us there. In fact, Revelations 21:2 speaks of our new home. "I saw the Holy City, the New Jerusalem, coming down out of heaven from God, prepared as a bride beautifully dressed for her husband." John goes on to describe the beauty of streets of gold, walls of jasper, gates of pearl, and foundations of precious jewels. Are you ready? Remember, Jesus said that unless a man is born again, he would never see heaven! Get ready, it won't be long!

"The heart is deceitful above all things." Jeremiah 17:9

In rummaging through the garage recently, I found a box filled with old school newspapers. One of them contained my picture, coupled with a brief interview. What struck me most was my response when asked about my greatest dislike. My answer was brief and to the point: "I can't stand phonies."

How profound. I might just as well have said I couldn't stand warm Coke or dents in my car door. Everybody dislikes phonies. Though there's fraud and deceit in all of us, it's just a lot easier to spot in other people. But who's fooling whom?

You misjudge a fly ball and it drops safely at your side. For the next half hour you cover yourself by telling everybody that you lost it in the sun.

You're in a grocery store and somebody waves. You wave back and smile, until you realize you've never seen the person before in your life. At that point you just pretend to be scratching your head.

Ten minutes after you arrive home from an evening out, your mother stops by your bedroom to ask how things went. You don't feel like rehashing the night, so you pretend you've fallen asleep.

A friend who's been sick asks you how church was on Sunday. You say it was OK–even though you slept in and watched a football game instead.

These are all little wrongs that don't really hurt anybody. That's probably what Ananias and Sapphira thought when they sold some of their property for the church, but kept some of the money back for themselves and reported a smaller selling price. Nobody else really knew the difference, but God knew of the hypocrisy, and struck them both dead on the spot.

In others we call such double-dealing hypocrisy. In ourselves, well, it pretty much goes overlooked, because we don't like to face up to the fact that, apart from God, we're all phony as a three dollar bill. Write down how you can be genuine today.

Have you ever been driving in a snow storm and lose control of your car? Being out of control is a very uncomfortable feeling. Your body tenses up, your grip tightens, and your heart rate increases. These symptoms remain true for a nation that is out of control. America is tense, the heartbeat nearing a fatal rate, yet the messages of righteousness, self control, and the after-life are rejected and unpopular.

Paul addressed these same unpopular subjects while on trial and was rejected much the way Christians are today. Acts 24:25 reads, *"As Paul discoursed on righteousness, self control, and the judgement to come, Felix was afraid and said, 'That's enough for now! You may leave. When I find it convenient, I will send for you.'"* Felix became very uncomfortable, so he developed a formula to deal with Paul. Our society has done the same. The message of the Gospel causes people to become afraid and uncomfortable. A gospel of convenience is created. "Get out of my face. I will send for you and your Jesus when I find it convenient." Friend, Paul never stopped speaking, and neither can we!

Everyone likes gifts, but most don't like problems. God will take both for you today.

Do you often struggle to work through your own problems and forget to give them to God? Have you learned to be sensitive enough to turn to God with all of your worries each day?

Read Jeremiah 33:1-3. Begin memorizing verse 3. This verse is often referred to as "God's telephone number" (J-E-R-33-3). in our rushed life-style, it is so easy to not slow down enough to really see what God's way is. We would rather try things our own way, which usually creates a lot of unnecessary worries. "Worry" is taking the responsibility for something God never intended for us to. If we can just learn to turn each day over to God, he will often show us great things far beyond our imagination.

Pause for a minute, and search yourself to see what it is you need to give to God today. Ask God to bless you in this area and show you how to deal with it.

JUNE 9   A CROWN THAT LASTS

In 1 Corinthians 9:25, an analogy is presented of the Christian life being like the Olympics. It says *"Everyone who competes in the games goes into strict training. They do it to get a crown that will last forever."* Athletes subject themselves to various methods of training, everything from waking up at ungodly hours to eating weird foods. Sort of like Sylvester Stallone in the original "Rocky" movie. He dragged himself out of bed at 5 AM, choked down five raw eggs in a glass, and started running. Most college students are just going to bed at 5 AM!

The Christian life does require strict training, as 1 Corinthians teaches. However, we are not competing for gold, silver, or bronze, but for an eternal crown that will be awarded to us by God Almighty. What a prize! You do not have to swallow raw eggs, but train yourself in God's Word daily, and exercise your faith as you run the race.

JUNE 10   I CAN'T HEAR YOU!

In the show "Gomer Pyle," the sergeant always screamed at Gomer, "I can't hear you!", only to have the battalion shout louder!

God usually speaks in a soft way, but we are often so busy that we fail to hear him.

Read 1 Kings 19:1-12.

Notice how God spoke to Elijah–in a still, small voice.

An insurance salesman called a house one day, and a small boy answered, in a whisper, "Hello?"

The salesman asked to speak to his parents. The boy responded, in a whisper, "They are busy."

The salesman asked if someone else was available to talk on the phone. The boy answered, in a whisper, "A fireman and a policeman are here." The salesman then asked to speak to one of them. The boy responded, in a whisper, "They are busy, too."

"Doing what," asked the salesman. "Looking for me," the boy replied and hung up the phone.

Sometimes, God is calling us in a whisper, but we are too busy to hear His voice. Will you slow down and listen for God's voice? Find out what God is quietly telling you.

JUNE 11   A NEEDLE IN A HAYSTACK

On its maiden voyage April 14, 1912, the cruise ship Titanic sank in the North Atlantic, some 350 miles off the coast of Newfoundland. Decades later, a man named Robert Ballard was driven by a life-long dream of one day finding this "needle in a haystack." For 13 years, Ballard searched the dark waters of the North Atlantic, and on September 1, 1985, the bright probe light and camera sighted the hull of this massive ship some 2 miles below the surface of the ocean.

Proverbs tells us that without a vision, people perish. Paul encourages us in Philippians 3:14 to aim at something, to have vision and goals in life. What are your goals in life? Do you have goals? Are they self-serving or God-serving? What are you doing to reach and attain your goals? With God's guidance and help, you can set reachable goals for your life that will enable you to be the person God designed you to be.

Read Deuteronomy 32:46-47.

Anorexia–avoidance of food–is an illness that sometimes affect young women. Steve Farrar's book, *Point Man*, tells the story of Karen Carpenter, a famous pop singer in the '70's who died of anorexia. She and her brother, known as "The Carpenters", recorded music that is still heard today on classic hit radio stations. At the peak of her career, Karen sang a number one hit song with the line, "I'm on the top of the world looking down on creation." She really was on top of the world at that time, making a tremendous amount of money and enjoying success in life. She could have eaten the finest foods from anywhere in the world, yet she died of malnutrition. What a tragedy.

In a spiritual sense, this happens to many students and young adults who avoid God's Word. God says His Word to us is life. Just as Karen Carpenter had available to her life's finest foods, so we have available to us God's words. Are you overcome by spiritual anorexia? It's there, but you must eat! Ask God to show you how to study His Word today.

JUNE 13   *THE BEAVER CLEAVER SYNDROME*

Do you remember when you were nine or ten years old and all you wanted in life was to be "grown up?" You and your friends couldn't go anywhere outside of your neighborhood because you couldn't drive. You couldn't stay up late because you "weren't old enough." As if that were not bad enough, older teens were always telling you to grow up! It sure was nice when you finally had the driver's license, and when you could stay up and watch the "Tonight Show." The benefits of growing older are nice, but along with maturity comes responsibility. Maybe that is why Paul says in Ephesians 4:15 to *"grow up into him who is the head, that is, Christ."* We must strive for the benefits of maturing, of being "grown up" in the Lord.

JUNE 14   R U GONNA STAND UP?

Read Daniel 3:1-18.

The Duke of Wellington was once out riding with a hunting party when he came to a gate guarded by a small boy. The lad said, "I'm sorry, sir, but my father told me to say that no one may hunt on his grounds." "Do you know who I am ?" the Duke asked roughly. "No, sir," answered the boy. "I am the Duke of Wellington," the man replied. The boy quickly took off his cap and looked toward the ground. "The Duke of Wellington will not ask me to disobey my fathers orders," he quietly said. The man then took off his hat and said, "I honor the boy who is faithful to his duty." The hunting party then turned and rode away.

The road less traveled is the one requiring sacrifice and commitment to values held high. How faithful are you to God's wishes for your life? Peer pressure is so strong. Like the boy and those faithful men who braved a hot furnace, do you stand strong in the face of pressure? Write down areas where you can stand strong.

JUNE 15   SUPER BOWL CHRISTIANS

Moments before Super Bowl II, Vince Lombardi said these words to the soon-to-be 1967 World Champion Green Bay Packers, "Just hit. Just run. Just block and just tackle. If you do that, there's no question what the answer's going to be in this ball game." These words contain a very important message for us today, because so many times we try to complicate things that should be simple. In other words, stick to the basics as Coach Lombardi was saying. Jesus said in Matthew 7:7 to ask seek, and knock, and if we will do this we will receive, find and the door will be opened. Jesus is urging us every day to stick to the basics. Just pray. Just read God's Word. Just put God first. Just follow Christ's example. If we do that, there's no question what the answer is going to be in this life.

JUNE 16   THE SEPARATION OF CHURCH AND SELF

Read Joshua 24:14-16.

Hypocrisy is a mega turn-off to everyone. Many Christians seem to wear two hats. One hat is their "holy hat," which they wear to church, and the other is their "worldly hat," which they wear the rest of the week.

The Kansas City Royals played the St. Louis Cardinals in the 1985 World Series. Since both teams were from Missouri, the Governor of the state was torn as to which team to cheer for. His solution? He had a cap made with Kansas City on one side and St. Louis on the other. When the Cardinals were up to bat, the St. Louis bill faced forward. When the Royals were up to bat, the Kansas City bill was forward.

In Christian walk, we cannot afford to wear God's hat and the hat of the world. We must choose whom we are going to serve. God doesn't put up with half-heartedness.

JUNE 17   PET ROCK PRINCIPLES

Read 1 Corinthians 1:27. It talks about "foolish things," something all of us can relate to. Like neckties. They would be very useful for wiping your mouth at dinner or blowing your nose, but, as it is, you pay $15-$20 for one, and there is no way you're going to wipe snot on it after that! Basically, they serve no real purpose. Remember the pet rock? Or mood rings? How about the Hoo-la-hoop or the Chia-Pet? It's all foolishness.

Have you ever stopped to realize that basic, moral, Biblical values are acclaimed as foolishness in the world's view? Those who subscribe to such values are considered foolish? How exciting, then, to read that *"God chose the foolish things of the world to shame the wise.."* Unlike neckties or pet rocks, you have a purpose and God will fulfill that in you every day. The world may call you foolish, but God is using you to confound the wise and strong!

JUNE 18   DIDN'T DO IT!

Read Genesis 4:1-9.

When accusation comes, we often want to pass the buck by saying, "I didn't do it." You know the story of Cain and Abel. Cain was certainly not close to God. When God did speak to him, he would either not listen or make excuses to avoid responsibility. His heart was cold, and he had to pay a price for it.

Satan tries to make us blame someone else or rationalize when God is speaking to us. We are tempted to admit to God only the easy things. God wants us to give him the areas in our life that need improvement.

Cain spoke evil against his brother Abel. We need to bless and encourage those around us, rather than tear them down! When God speaks, if we will listen, we will not feel the need to pass the buck.

JUNE 19   THE INTIMIDATOR

On a Saturday visit to the zoo, a group stopped to view the "King of the Beasts," as the lion is known. He was lazily snoozing in the warm sun, with his female counterpart lying close to his side. After a few moments, the lioness stood up and walked away, disappearing into a cave. All of a sudden, the male lion lifted his head, discovered her absence, and let out a roar that could be heard all over the park. The lioness immediately walked out of the cave and laid down in the exact spot she was before.

In 1 Peter 5:8, we read, *"Be self-controlled and alert. Your enemy, the devil, prowls around like a roaring lion looking for someone to devour."* Satan is like a lion whose teeth have been pulled. All he can do is roar loudly to try and intimidate you. He will succeed in doing so if you fail to do what the first part of this scripture says, be self-controlled and alert! Stay on your toes. When the enemy roars, remind yourself that he is a toothless foe.

JUNE 20   WHICH WAY DO I GO?

Read Jeremiah 29:11

Have you ever sat at a traffic light not knowing which way to go?

When Adoniram Judson graduated from seminary, he received a call from a large church in Boston to become an assistant pastor. He received many congratulations, along with advice from people who tried to persuade him that this was a once-in-a-lifetime opportunity. His widowed mother and sister also rejoiced that he could live at home with them. Judson refused the call, saying, "My work is not here. God is calling me beyond the sea. To stay here, even to serve God in this ministry, I feel would be only partial obedience, and I could not be happy in that." Although it was a great struggle, he left his mother and sister to follow God's call. Judson's church in Burma had 50,000 converts, and the influence of his dedication has been felt around the world.

Have you sensed what God wants you to do in life? His plan for you will always be the best! Write down what direction you should go to follow God and be obedient to follow it.

Take a look at these words from an old song by the group "Mickey & Becky":

> "Well, there's not much to tell whenever I'm compelled
> to share my testimony,
> No hair raising tales of hell-raising days
> With the demons dancing all over me.
> Well, I never smoked dope or swore at the Pope
> Or spent the night with a shady lady.
> I just came in bold when I was 6 years old
> And said, 'Preacher, I want Him to save me.'"

What is your testimony? Maybe, you don't feel you have one, or at least not one that is exciting. Revelations 12:11 says, *"They overcame him (Satan) by the blood of the Lamb and by the word of their testimony."* Maybe God saved you out of a life of crime or hard drugs. Or maybe, like the song, you gave your heart to Jesus at an early age and remained faithful to Him. Whatever the case, you can defeat Satan daily by sharing your testimony and giving praise to God.

Read 1 Samuel 3:8-10.

American Indians would sometimes lie silently with their ears to the ground to hear whether unseen horses of their enemies were galloping toward them. The Indians could not hear their approach when they were up and moving. This meant their lives could be in danger, because they did not know in advance that their enemy was coming.

If we can't learn to sit quietly at the feet of Jesus, we might miss instructions that are necessary for our spiritual safety.

Eli told Samuel to go and lie down and listen to the Lord. It is so hard to squeeze quiet minutes in with the Lord in the busy schedules most keep today. However, to hear God's voice, we must find a place of quiet to hear Him speak.

Take a few minutes today at lunch or before class and sit listening to God. Ask Him to speak to you. Do it now!

JUNE 23   THE STUFF PEOPLE ARE MADE OF

Most people in the world want to be people of character, which is defined as "the total quality of a person's behavior," or basically, what you're made of. D. L. Moody said, "Character is what you are in the dark." Robert Freeman made this comment, "Character is not made in a crisis—it is only exhibited." So how can we become strong in character?

Are you ready for this? Character is produced through suffering. "Not me" some would say. "I don't do suffering." Romans 5:3 says *"...we also rejoice in our sufferings, because we know that suffering produces perseverance; perseverance, character; and character, hope."* Not a popular concept in our "me" generation, but a Biblical concept. There is definitely a price to pay to be a person of character. As Horace once said, "Fame is a vapor; popularity an accident. Riches take wing; those who cheer today will curse tomorrow; only one thing endures—character."

JUNE 24   PROTECT YOURSELF!

Read Psalm 91:1-4.

The safe sex ideology is "protect yourself." It is not a fool proof of protection.

A Christian man was fleeing from his enemies during a time of persecution in North Africa. Pursued over a hill and through a valley with no place to hide, he fell exhausted into a cave, expecting to be caught. Awaiting his death, he saw a spider weaving a beautiful web across the entrance to the cave. When the man's pursuers arrived and saw the unbroken web, they assumed it was impossible for the man to have entered into the cave. Later, the man explained "Where God is a spider web is like a wall! Where God is not, a wall is like a spider's web."

The greatest security for believers is an awareness that God will use His power to take care of them, even when the odds against them seem overwhelming. Have you ever been in a situation where you saw God's protection first hand? Are you going through something now where you need it? Write down how God has protected or will protect you.

JUNE 25   THE PRAYER OF FAITH

A young minister headed straight for the hospital upon arriving in town in order to visit his grandmother. He was met at the door with news that her surgery had unexpectedly revealed organs covered with cancer. The doctors said there was nothing that could be done. Late that night, the young minister felt God speak to his heart. "Lay hands on your grandmother and pray for her," he heard God say. The bargaining began, "But God, I can just pray for her here," he said. "I don't know how she feels about divine healing."

The young minister knew God was a healer and believed God's promise in James 5:15, *"And the prayer offered in faith will make the sick person well."* He obeyed God and two days later received the news that the doctors could now find no sign of cancer.

Will we ever understand everything about divine healing? No, but that doesn't change the fact that God is our healer, a miracle worker, and that we should trust Him daily for healing and for miracles. Call out to Him in faith.

Read Proverbs 7:24-27.

All around us are caution signs–slippery when wet, high voltage, danger ahead, smoking can be hazardous to your health. We heed caution signs at all sides; except when it comes to God's words.

Proverbs 7:24-27 is a caution sign that boldly says to stop and turn around when confronted with temptation of fornication or adultery. If you don't, it will mean your downfall.

Age and treachery will overcome youth and zeal. Many college students use zeal to overcome an old and treacherous foe–the devil. What we need is wisdom! God's Word is wisdom to the seeker to give caution to potential pitfalls.

Have you ignored a warning from God's Word lately? Just as there is an important reason for every caution sign, so is there one for each of God's warnings. Ask God to show you again areas you need to pass with caution, and be obedient to His Word.

Read 2 Corinthians 5:17. These are powerful words written by Paul, a former murderer and specialist in persecuting Christians. He writes, *"Therefore, if anyone is in Christ, he is a new creation; the old has gone, the new has come."*

In 1970, Harold Morris found himself in the Georgia State Penitentiary with two life sentences after being falsely charged with murder and armed robbery. After accepting Jesus as his personal savior while in prison, God began to open doors for this talented man to speak to young people. His chances of ever leaving prison looked doubtful, but through much prayer, Harold Morris was granted parole in 1978. The best was yet to come. On May 15, 1981, one day before graduating from Bible college, he was granted full pardon. The crime against him was erased and his rights as a citizen restored.

Praise God for granting us a full pardon. All of the old has been erased and our rights as a citizen of heaven have been fully restored.

Read 1 Samuel 23:15-18.

It has been said that there are two types of people in the world—basement people and balcony people. Basement people are those who call to us from below and bring us down. They say things like, "You can't." "Who do you think you are?" "You'll never amount to anything." "Give up!" Balcony people on the other hand call to us from above and bring us up. They say things like, "You can!" "You're special." "Don't give up."

Do you know balcony or basement people? Which one are you?

David was down and afraid. He was doing his best to serve God and the king, but the king was out to kill him. It would have been easy for David to blame God for his trouble and turn his back on him. He could have thought, " If God cares for me, how could He let this happen?" Along came his friend Jonathan and *"helped him find strength in God."* Jonathan reminded David of God's love and His plan for David's future. Jonathan called from the balcony and pulled David up when he was down.

Do you pull up with encouragement, or push down with criticism or negativism? Pay close attention to what you say and how you say it. Whom you can pull up today? Look up for God to lift you higher!

Time is so often our worst enemy. We intend to do certain things, but just can't seem to find the time. Before we know it, days, weeks or even months have spilled by and we still have not done what we need to do.

An old Iowa couple were eating dinner when a fierce tornado hit their home and tore the roof off. The wind picked the couple and the table up, carried them one mile, and set them down in the middle of a corn field. "Why are you crying, dear?" asked the husband. "We are okay."

"I know," the wife replied, "but it is the first time we have been out to eat in years and I'm just so happy."

Christ longs to spend quality time with us, but if we are not careful, we will find time for everything else except Him. Ephesians 5:15-16 tells us that we must live as wise people, and we should redeem the time *"because the days are evil."*

JUNE 30   NICKELS, PENNIES & DIMES—IT'S ALL GOD NEEDS

Read Daniel 5:1-5, 22-30.

Belshazzar reigned after King Nebuchadnezzar in Daniel's time. Both of these kings were very proud. God had to humble them and break their pride. Nebuchadnezzar changed after his humbling experience, but Belshazzar did not. He was killed!

God wants to change our lives. The evidence of a Christian is a changed life-style. If Christ is a part of your life or living inside of you, it's bound to show.

Belshazzar thought he could totally depend on things rather than God. He never really knew God. He continually sought to bring glory to himself.

The more potential you have for success gives more potential for pride. We need to learn to give God all the glory! What do you need to give God credit for in your life?

Read Amos 5:14-15.

Living in an amoral relativistic society can be pretty scary sometimes. Amos was a prophet who preached to people at a time when there was great wealth and prosperity, and people thought they were secure. Their religion was very unstable, because God was not a part of their lives. Amos was very bold and courageous as he stood for God and warned the people that prosperity would not last forever.

We need to be as serious about standing for what is right as Amos was. We need to hate evil and hate sin. We need a strong love for what is good and boldness and courage to stand for those principles. If we just sit back and enjoy the easy life, Satan wins, we lose. Whom are you for?

Pray that God will begin to build a boldness in you and give you the strength to speak out when you know something is wrong, even if you know it may cost you. How can you stand up today?

Read 2 Timothy 4:1-4.

A favorite pastime of America's college students is going to the movies. It's been an American pastime for decades, but the Bible talked about it thousands of years ago. Today's Hollyweird (oops Hollywood) is moving at light year speed away from traditional values that shaped our great nation—family, prayer, community, social responsibilities, and heterosexual, monogamous relationships. What we see has a real strong impact on how we live!

Ted Bundy, a convicted and executed murderer, confessed in 1989 to Dr. James Dobson, Family Psychologist, that entertainment and pornography shaped his thoughts so much that it drove him to murder 28 times.

The Bible says in 2 Timothy that, in these last days, we would be turned aside to fables. The infatuation of today's media with fantasy and mythology is proof it's happening right now! No action is an action before it is first thought. What's showing in the theater of your mind? What are you feeding your thoughts? Psalm 1:1-5 says those who meditate on God's ways and words are like a tree firmly planted!

Write down how your mind can think and show the things of God's ways. Meditate on Jesus—do it now!

JULY 3   LAUNDRY LESSONS

A freshman from the University of Michigan carried a huge load of laundry to the laundromat for his first washing experience. Dumping all of the clothes onto a wash table, he began separating them into three piles—lights, darks and whites. Mom's laundry lesson echoed in his head. Suddenly, Romans 10:12 came to mind. He scooped up all three piles and threw them into one washer together.

Racism is a prominent issue in our society. However, as Romans 10:12 states, *"...there is no difference between Jew and Gentile—the same Lord is Lord of all and richly blesses all who call on him."* There is only one heaven where we will all spend eternity together, but God's mandate for us while here on Earth is to love one another with His love and together, build His kingdom.

JULY 4   TRUTH OR CONSEQUENCES?

Do you believe in absolute truth? The majority of adults (66%) do not. They believe different people can define truth in conflicting ways and still be correct. Of born–again Christians, 23% believe there is no such thing as absolute truth. Of college–age youth and young adults, 30% believe there is no such thing as absolute truth.

Our nation was founded and built on the absolute truth of God's Word. But as more and more people dismiss absolute truth, which dismisses the principles of God's Word, the foundation of our nation slowly erodes. Absolute truth does not represent bondage as so many are inclined to believe. It represents freedom. It is like a young teenager wanting to be free from his parents' rules so he can smoke and drink. That cigarette or beer represents freedom to him. Once he gains that "freedom," he soon finds himself in bondage to nicotine or alcohol. His parents' rules were meant to protect. God's truth protects us and keeps us from bondage. John 8:32 says, *"Then you will know the truth, and the truth will set you free."*

JULY 5   THAT'S MINE!

Have you ever heard that?

Isn't it funny how a $5 bill looks like so much when you drop it in the offering plate, but so little when you go out on Friday night.

It is estimated that the buying power of American college students is in the tens of billions per year. That's a lot of money filtering through the hands of college students. How much of it makes it to God?

"Tithing, oh that's for adults!" That's just not true. It all belongs to God—even your allowance. All he wants is one-tenth of it in return—you keep the rest. If you ask me, that's a pretty good deal.

Read Malachi 3:10. Commit to start tithing this week. Write down the amount of one-tenth of your income. God's already kept His end of the bargain. Why don't you? Ask God what He wants you to give today.

JULY 6   MAN GIVES BIRTH; SEE PAGE 5

Can you imagine a man having a baby? Not! Men will never know the agony of childbirth, but we, both men and women, are told to reproduce the life that is in each of us. Read Genesis 1:28.

It is said that no two snowflakes, no two fingerprints, no two grains of sand are alike and no one else is like you. You are unique! Yet, at the same time, we are told to multiply the earth and reproduce the life of God within us.

David, on his death bed, wanted to reproduce in Solomon his secrets of success. David told Solomon:

1. Act like a man of God.
2. Be true to the word of God.
3. Rely on the promises of God.
4. Execute the judgements of God.

What are you reproducing in others? Potential for greatness or findings of failure? Write down how you can reproduce spiritual life in someone else's heart today.

JULY 7   LAUNCH OUT

Webster's dictionary defines faith as trust or confidence. Naturally, for the Christian this means trust or confidence in God, yet so many people struggle with the issue of faith verses logic. Everything must be explained and understood. If it cannot be, then it must not be real. Faith (trust and confidence) is an essential element in any relationship. Someone has said that you cannot discover new oceans unless you have the courage to lose sight of the shore. We may not see every detail of the Christian life unfold before us, but we must learn to walk by faith, not by sight. Hebrews 11:1 says, *"Now faith is being sure of what we hope for and certain of what we do not see."* If you are facing a major decision, or if you are at a crossroads of some sort, trust God. Launch out and walk in the confidence that God will guide your every step.

Marijuana use on college campuses was up sharply in 1993 from the previous year. A popular phrase in drug users' vernacular is "Hey man, let's fire one up!" In 2 Chronicles, we are told to F.I.R.E. One up–to God!

Fervent
Intercessors for
Rapid
Evangelism

What's wrong in America is not the devil's fault. It's the church's fault. The Bible says healing comes when we begin to pray. Prayer puts us in touch with the immediate resources of heaven. However, we get so busy, many times we neglect to talk to Jesus about our needs. How is your prayer life? Are you firing off signals to God? Stop now and pray (Matthew 6:9-14) and ask God how you can fire someone up for him? A prayer today keeps heart decay away!

Probably one of the most often quoted scriptures in the Bible is Romans 8:28. It says, *"And we know that in all things God works for the good of those who love him, who have been called according to his purpose."* This does not mean that only good things will happen to us or that our lives will be trouble free. The comfort comes in knowing that if God is top priority in our lives and we are serving and loving him with all our heart, no matter what happens to us, God will cause it to work for our benefit.

Dave Roever was standing on a boat with a grenade in his hand during the Vietnam War when all of a sudden, a sniper's bullet hit the phosphorus grenade and exploded it inches away from Roever's head. The phosphorus continued to burn even while he was floating in the water. Roever sustained burns over much of his body. Now, with a disfigured face and hands, Dave Roever travels all over

America, telling how God brought him from tragedy to triumph. He is Romans 8:28 in action.

JULY 10   I THINK I CAN, I THINK I CAN...

Read Nehemiah 4:4-6.

Are you ever afraid others will make fun of you for living the Christian life? Are you aggressively accomplishing things for God today?

Many people made fun of Nehemiah and his people as they attempted to rebuild the walls of Jerusalem. They didn't think they could do it. The scripture says the walls were soon half way built because the people were eager to work for God.

I once saw a poster that said, "All I want is less to do, more time to do it, and more pay for not getting it done." This is often a big time collegiate attitude. Plan to accomplish twice as much today as you normally would. If you are serious, you will probably finish in half the time! Write down something you've been putting off and accomplish it today.

JULY 11   MOTHERLY LOVE

In the late 1940's, a Catholic nun began ministering to the poor people of Calcutta, India. Mother Teresa would continue her work for many decades, helping to start ministries in over 71 countries around the world where her Missionaries of Charity would work among the poorest of the poor. Mother Teresa, who won the Nobel Peace Prize in 1979, said, "In serving the poorest, we are directly serving God."

Jesus said something very similar almost 2000 years ago in the parable of the sheep and the goats. It is found in Matthew 25:40. He says, *"The King will reply, 'I tell you the truth, whatever you did for one of the least of these brothers of mine, you did for me.'"* It may not be convenient, but to touch the life of someone in need is to touch the heart of the Father. Take time, make time, to minister to the hurting.

JULY 12   EVERYONE SINGS IN THE SHOWER

Everyone sings in the shower. The acoustics are great and we all feel like a star. When the showers of life come, we need to sing to God.

Enter his gates with thanksgiving and his courts with praise; give thanks to him and praise his name. For the Lord is good and his love endures forever; his faithfulness continues through all generations. Psalm 100:4-5

I can live for two months on a good compliment. Mark Twain

Praise God from whom all blessings flow; Praise him, all creatures here below; Praise him above, ye heavenly host; Praise Father, Son and Holy Ghost. Thomas Ken

One thing scientists have discovered is that often-praised children become more intelligent than often-blamed ones. There's a creative element in praise. Thomas Dreier

Praise makes good men better and bad men worse. Thomas Fuller

Are you focusing on praise or problems? Sing a song to God. He promises a quick shower of His blessing of protection and provision.

JULY 13   A POINT OF FOCUS

While on earth, Jesus had an incredible ability to relate to people from all walks of life. In Luke 9:62, Jesus makes this statement while conversing with a group of men along the road, *"No one who puts his hand to the plow and looks back is fit for service in the kingdom of God."* What are your goals today? What is your focus as far as your walk with Christ is concerned?

Craig was a young, energetic farmer's son who was excited about helping on the family's Texas farm. Now eleven, his dad had finally agreed to let Craig drive the John Deere tractor and plow the field. Craig began the first row and carefully watched as the plow turned up the ground, but upon completing the first row, he noticed the row was S-shaped, like a snake. "What did I do wrong," he asked his dad. "Son, when plowing, you must pick a point straight ahead in the distance and drive straight towards that point. Never look back at the plow." Jesus desires the same from us, that we fix our eyes on the goal and never look back.

JULY 14   CHARACTER DOESN'T COUNT

In the 1992 presidential election, it was said that character was not a political issue. However, to God, it has always been of supreme importance.

A good name is more desirable than great riches. Proverbs 22:1

Character is what you are in the dark. Dwight L. Moody

Reputation is what folks think you are. Personality is what you seem to be. Character is what you really are. Alfred Armand Montapert

Character is not made in a crisis—it is only exhibited. Robert Freeman

What do others perceive your character to be? Write down how you can establish God's character in your life.

JULY 15   THE HERO OF HEROES

All of us at sometime have heroes, people we pattern our lives after and emulate. Athletes, comic book stars, business leaders to name a few. The person we should strive most to imitate, the greatest hero whoever lived, is Christ.

He experienced everything we face today while he walked on

earth. Jesus was tempted. He faced difficult trials, experienced crises, and knew the reality of death.

In the book, *In His Steps*, by Dr. Charles Sheldon, we find the story of how Rev. Henry Maxwell challenged the members of First Church to base every decision on one simple question, "What would Jesus do?" Ask yourself that simple question when faced with trials, pressures or temptations, and imitate Christ.

JULY 16   *WOODLAND HILLS TRAGEDY*

Read Exodus 20:13.

On February 2, 1982, workmen repossessed a boxcar-sized dumpster from the yard of Malvin R. Weisberg of Woodland Hills, California. When the padlocked doors were opened, workers faced a wall of wet boxes and a stench that was thick as fog.

As the cartons were unloaded, the bottom of one suddenly tore, and clear plastic containers tumbled to the floor. When the workers gathered around to examine the contents, they were horrified. "It was human flesh, pure and simple," said Ron Guilette, yard supervisor.

Eventually, 16,433 bodies were retrieved from the storage container. Each was a baby who died before birth. The cause of death: abortion.

The tragedy is that children are aborted nationwide, three to four thousand a day. In the next twelve months, nearly 1.5 million unborn babies will lose their lives in this country alone. To compound the tragedy, each year in the United States approximately 1.6 million requests for adoptions aren't met because there aren't enough babies to go around.

We are all familiar with the story of the "unsinkable" Titanic that hit an iceberg on its maiden voyage April 14, 1912 and sank. Many of us are probably not aware of the fact that the ocean liner Californian, only nine miles away, watched this tragedy at sea take place. The Titanic shot up distress signals, seen by the Californian, but misinterpreted. The Titanic was sending radio signals, but the radio operator for the Californian was asleep. Sadly enough, many people all around us are sending out distress signals, cries for help, yet in our hectic lives, we sleep and are unaware of the numerous tragedies that occur. Paul says in 1 Thessalonians 5:6, *"...let us not be like others, who are asleep, but let us be alert and self-controlled."* Make it a point to be sensitive to those around you who may be in need.

Some people believe we should look the other way when it comes to abortion, if only because this is America. It's a free country, right, and nobody's supposed to have the right to impose religious beliefs on another person? True, but it's also a country wherein nobody is granted the right to deprive another person of life and liberty. To do so is not imposing a religious belief, but violating a principle that's basic to every culture in the world: *Thou shalt not kill.*

Murder may not only be physical; it can be of a relationship. How can you avoid murder today?

Compared to all the big-time, high-profile saints and scoundrels who show up in the Bible, Jabez is a small fry. Placed side by side with the prophets, potentates and assorted nabobs, he's a nobody.

All that's known about him is conveyed in 1 Chronicles 4:9-10. He acquired his name (which sounds like the Hebrew word for pain) because his birth practically did his mother in. Although he was a good, clean-living kid, he seems to have had a rough life–perhaps because other kids teased him about his name.

But the day came when he decided enough was enough, and he asked God to help keep him safe and to give him bigger challenges– to *"enlarge his territory."* God answered his prayer.

I believe God is eager to do the same for any one of us, providing our ambition is not motivated by selfish desire. The problem is, we're often too content in our own little ruts. It's never crossed our minds to pray that God would enlarge our horizons.

How can you trust God to meet your requests?

JULY 19   FOLLOW THE LEADER

In the Gospel of Matthew, we find recorded the account of the wise men who came to Jerusalem to see Jesus. These men were led to where Jesus was by a bright star, a light in the darkened sky. In John 8:12, some years later, Jesus said of himself, *"I am the light of the world, he who follows me shall not walk in the darkness, but shall have the light of life."* Wise men today are still following the light in the darkened sky, but the light is not leading them to Jesus. Many people however, are unwise and are following false light. If you are tired of the darkness of your life, let the light of Jesus flood your soul today and give you the light of life.

JULY 20   EXPERIENCED ONLY

Read Zechariah 4:6.

Looking for a job, Rebecca opened the Sunday paper and flipped through the want ads. The little boxes shouted at her, "Experienced only!"

Sometimes it may seem that the world is conspiring against you. At a time when you feel very adult and deal regularly with adult-sized concerns, you are constantly belittled and reminded of your youth.

Jeremiah complained he wasn't old enough to do what God had in mind for him. But he was promptly reminded by the Lord, *"Do not say, 'I am only a child'... For I am with you."* Physical maturity matters less than spiritual maturity, regardless of the world's standards.

Then as now, God can use your life in extraordinary ways if you trust Him in very ordinary ways everyday. Forget about age and talent and limitations. Put aside your feelings of insecurity and inexperience. For the Lord has big plans for those who care about their stature in God's eyes more than their stature in man's eyes.

Write your career, personal and spiritual goals down. Pray for God's direction.

JULY 21   EASY FOR YOU TO SAY

Two of the most common words in the English language are the words "thank you". Why is it that those two words are so difficult to utter at times? In Luke 17, Jesus meets ten men who have leprosy, a disease we might equate with today's AIDS virus. These men beg Jesus for mercy, and He heals them. One man came back to Jesus, fell down and thanked Him repeatedly. In verse 17, Jesus asks, *"Where are the other nine?"*

If you have ever done something for someone and received no thanks, you know it can hurt your feelings. Saying thanks to a roommate, parent, or friend takes little effort but goes a long way. Even more important than that, we should make a conscious effort everyday to say thanks to Jesus for His overwhelming blessings and love.

JULY 22   SOUNDS OF SILENCE

Read Zechariah 2:13.

In the middle of the Indiana woods sits a tiny hut constructed of rubble scrounged from nearby building sites. Resembling a tool shed more than anything habitable, it represents the major turning point in the life of John Michael Talbot. He built the shack after everything in his life fell apart. Alone, in solitude, he sought God.

At first, he had romantic thoughts about living as a hermit and being "spiritual." Before long he began to experience boredom. He discovered the true meaning of loneliness.

"I'd read the scriptures and think, 'I've read this before. I don't need to read it again,' said Talbot. "But it was during these basic, boring days that I experienced a deeper sense of just being. And in my silence I heard God's most profound words."

John Michael admitted that people aren't going to rush out and build Daniel Boone huts in the woods just to experience some spiritual dimension of silence. But he also said they don't have to. They can easily experience it where they are.

Stop now and listen to God's voice, so that you can live for God.

JULY 23   HEY BUD, WISE UP!

"Everything is permissible for me, but not everything is beneficial."
These words were spoken by Paul in 1 Corinthians 6:12 in reference to the excessive sinful behavior of the Corinthians.

Drinking among college students is on the rise in America. Recent studies have shown that 45% of college students use alcohol on a weekly or more frequent basis. Some 42% of the students surveyed consumed five or more drinks in one sitting within the last two weeks.

The Swedish bikini team, Bud Bowl, beach parties, and tropical paradises have made drinking all the more glamorous in the eyes of America's young adults.

Determine today what is right and beneficial for you based on the standard of God's Word. Stand tall in your decision, knowing that God will honor you for obeying His Word.

Christina was raised by her mother in a poor Brazilian village. Her dad died when she was an infant. As a teenager, Christina grew tired of the uneventful life on the village and constantly wanted to go to the city. The very thought of this horrified her mother, Maria. Maria knew what her daughter would do or would have to do for a living. That is why she was broken-hearted one morning when she awoke to find Christina's bed empty. She quickly threw some clothes in a bag, gathered up all her money and ran out of the house. On her way to the bus stop she took time for one last thing— pictures. With all the money she could spare, she took picture after picture of herself. When Maria arrived in the city, she went to the bars, hotels, any place with a reputation for prostitutes. At each place, she left her picture with a short note on the back—taped on a bathroom mirror, tacked to a hotel bulletin board, fastened to a corner phone booth. When the pictures ran out, Maria sadly left for home.

A few weeks later, Christina walked down the stairs of a hotel, exhausted and filled with pain and fear. As she reached the bottom of the stairs, the lonely girl saw a familiar face. There on the lobby mirror was a small picture of her mother. She slowly walked across the room and removed the photo. On the back was a note from Mom. "Whatever you've done, whatever you've become, it does not matter. Please come home." She did!

God has done the same thing in His Son, Jesus Christ. He is the picture of God saying to you, "Whatever you've done, whatever you've become, it does not matter. Please come home." Read Jeremiah 31:3.

JULY 25   HIGH TIDE

Ah, the joys of college life, freedom, new friends, guys and girls, sports, laundry. Laundry? Oh those piles of filthy clothes that greet you by sight and smell each time you walk in the door. Do you feel the "tide" mounting? Don't "shout" at your roommate. "Cheer" up, there is great "gain" in clean clothes. What a relief to finally throw the last load in the dryer, to stretch yourself out on clean sheets or to pull on a fresh, clean shirt.

Remember how good it felt to invite Jesus into your heart and be totally clean on the inside? What a relief! New life and energy surges through you as the old self dies and newness begins. In Acts 10:15, the Bible says to *"not call anything impure that God has made clean."* In other words, the washing of the blood of Christ is total and complete. We are pure and clean, and the old smell of sin is replaced with the sweet smell of salvation. Is there sin or filth in your life? Let the detergent of Christ's blood wash you today and make you whole and clean.

JULY 26   THE EFFECTS OF SIN

Read Proverbs 20:7.

Harold Morris tells a true story about a car accident in which a teenage boy was killed. He and his friends had been drinking. When the police reached the home of the deceased boy and explained the details of his death, his father's immediate reaction was, "If I could find who sold those boys that alcohol, I'd kill him!" The dad was so shaken up that he went to his liquor cabinet for a drink and there he found a note that read, "Dear Dad, I knew you wouldn't mind if we borrowed this bottle of liquor. We'll pay you back soon."

The father thought that his drinking affected only his own life, but he was sadly mistaken. His life of sin was soon copied by his son. How tragic the truth is. No one sins just to himself, his sin influences others around him.

Whom is your sin affecting? Someone in your family? Is it worth it?

Do you ever wonder why you keep making the same dumb mistakes over and over? Well you are not alone. A college coed informed her suitemate one morning that she had started a fire in her microwave. She explained that it took so many paper towels to absorb the grease when cooking bacon that she thought she would save money be using newspaper. "Did you ruin the oven?" Her suitemate asked . "Well," she answered reluctantly, "not the first time..."

Why do we keep giving in to the same temptations or failing to make the right choices? Paul struggled with this "flesh vs. spirit" war and said in Romans 7:15, *"I do not understand what I do. For what I want to do I do not do, but what I hate I do."* So does that mean we should give up and give in to sin? Absolutely not! We must learn from our mistakes and determine in out hearts to crucify the flesh and live a godly life.

How much do you feel you need your Christian brothers and sisters? Read Ecclesiastes 4:9-12.

One of Aesop's famous stories relates that a farmer had sons who were always quarreling among themselves. One day the farmer asked his sons to gather up sticks and place them before him. Having tied the sticks in a tight bundle, the father ordered each of his sons to try to break the bundle. They all tried, but none was strong enough. Then, untying the bundle of sticks, the farmer handed his sons the sticks to break one by one, which they did easily. The wise farmer said, "Thus, my sons, as long as you remain united, you are a match for all your enemies, but differ and separate and you will be destroyed."

You know, we in the Church really need each other. As

Ecclesiastes says, in numbers there is help, comfort and strength. Write down two people in your church for whom you are thankful. Then give them a call or write them and let them know you appreciate the strength their friendship gives you.

JULY 29   HOW TIME FLIES...

Read 2 Timothy 2:15. Okay, maybe not the most popular scripture for a college student, because it says, *"Study to show thyself approved unto God..."* Someone has said that in an average life of seventy years, time would be used somewhat as follows:

3 years in education
8 years in amusement
6 years in eating
4 years in conversation
5 years in transportation
14 years in work
3 years in reading
24 years in sleeping
3 years in convalescence

But how much time does a man give God? If he attended a church service every Sunday for 70 years, prayed 5 minutes each morning and evening, spent 30 minutes every week in Bible study, he would be giving 13 months. 13 months out of 70 years.

JULY 30   SCARED STIFF

We probably all know that sick feeling in the pit of our stomach spelled F-E-A-R. Some fears are healthy, like the fear you feel that keeps you from leaning too far over a cliff, but fears can be paralyzing. This would keep us from accomplishing the tasks God has given us in life. God does not want us to be immobilized by fear. Read Isaiah 41:9-13.

Several decades ago an artist named J.H. Morthian read that a tiny boy had been killed on a busy roadway. Fear gripped him that such a tragedy could happen to his own children. He became obsessed with providing them an atmosphere in which they could be totally safe. He bought a house in the country where there was virtually no traffic. He built the children a play area 50 feet from the driveway and built safety features into every part of his home. Ironically, before a turnaround driveway could be built, Morthian's 18 month old son was struck and killed as Morthian himself backed out of the carport.

Fear can be both unfruitful and fatal, but God can deliver us from unhealthy fears. He promises to give us help in whatever difficult or threatening situation we are facing. Write down the area in which you need His help and courage today.

JULY 31   SHARK ATTACK!

Okay, get ready for this. Everyday in the United States, 105 people become millionaires. Does that just totally encourage you? There you are, hanging out at the Student Center, hoping someone will drop some change just so you can do laundry, or praying Mom will send a care package with stamps in it. Luke 6:38 says if you give it will come back *"good measure, pressed down, shaken together and running over..."* However, that does not pertain to the lottery. Statistics show that a person is three and a half times more likely to be killed by lightning and five times more likely to be eaten by a shark than he is to win a state lottery. That scripture pertains to giving money, time or talent to the work of the Lord. God may not make you a millionaire today, but if you give to Him, He will bless you in ways you cannot imagine.

Everyone likes encouragement. It feels good to be told "you can do it" or wow, what a great job you did!" A daughter made a feeble attempt of encouraging her mother. The mother was discouraged, even though she had lost two pounds, about not losing it in the "right places." "Mom," the daughter spoke up, "you have an hourglass figure. It's just that the hour is getting late."

Hebrews 10:24 says we should *"consider how we may spur one another on toward love and good deeds."* Make it a habit in your life to find ways every day to encourage others. In the process, you will find yourself being encouraged!

This day I call heaven and earth as witnesses against you that I have set before you life and death, blessings and curses. Now choose life, so that you and your children may live and that you may love the Lord your God, listen to His voice and hold fast to Him. Deuteronomy 30:19-20

Of all of the words in the English language, my favorite is yes. What other word makes people more happy or offers more hope? What other word can bring a bigger or faster smile?

"Yes, the exam has been cancelled."

"Yes, the X-rays are clear."

"Yes, you're hired."

"Yes, I will marry you."

Yes, brings the good news. It seals the commitment. It gets rockets launched and new cars designed and new shopping malls built. Yes is affirmation that your are on the right track and you are doing a good job. If I knew I were going deaf and could actually choose the last word I ever heard before my ears ceased operating, that word would be yes.

Yes is at least as old as the words of Moses in Deuteronomy, *"This day, I have set before you life and death, blessings and curses. Now choose life."*

The first day of the rest of your life began when you said yes to God. That may have happened ten years ago. It may have occurred yesterday. On that day, whenever it was, you chose life. Eternal life.

Pause a few minutes. Think back to that special day and to the events leading to your decision. As you recollect, jot down a few of the details that come to mind. In addition, make a note of any changes in your life that you've noticed then.

AUGUST 3   A BIBLICAL BALANCE

Someone once said that God wisely designed the human body so that we can neither pat our own backs nor kick ourselves too easily. We live in the "me" generation where people are encouraged to magnify self and boast of achievements, but the Bible says in James 4:6 that *"God opposes the proud but gives grace tot he humble."* The One we should magnify is Christ, not self. God hates a prideful spirit, but neither does God want us to be so down on ourselves that we cannot accept His love. We must find a balance and never forget that we are nothing without God, but with Him and through Him, we are royalty.

AUGUST 4   A ROAD MAP FOR LIFE

Dr. George Washington Carver once said, "There is no need for anyone to be without direction in the midst of the perplexities of this life." Read Proverbs 3:5-8.

Dr. Carver used to get up early each morning for his quiet time alone with the Lord. In relating how important those early morning times were to him, he said, "At no other time have I so sharp an understanding of what God needs to do with me as in those hours when other folks are still asleep. Then I hear God best and learn His plan!"

When we spend time with the Lord in His word and in prayer, He gives us direction in our lives. He teaches us principles by which we can make good decisions and He makes us more sensitive to the nudge of the Holy Spirit.

Which area do you need God's direction this week? Plan to spend a few minutes every day talking to God about that area and asking Him to lead you in the direction you should take.

AUGUST 5   HANG IN THERE, HARLAN!

Do you ever feel like quitting? Read Galatians 6:9 and take a look at Harlan's life. At age fourteen, he dropped out of school, worked some odd jobs as a farmhand, but hated it. At 16, he lied about his age and joined the Army, but he quit after one year. He then tried blacksmithing but failed. He got on with Southern Railroad, got married, and then was fired from his job. Then came more railroad jobs. Harlan tried studying law, selling insurance, selling tires, running a ferry boat, and running a gas station. It was no use. He was a loser and now at age 65, he was reminded of that with his first social security check and nothing to show for his life. Did he quit then? No, Harlan took that check and opened a restaurant. The restaurant he started was Kentucky Fried Chicken. You know Harlan as Col. Harlan Sanders. Don't give up!

AUGUST 6   MERCY

Begin now and read Lamentations 3:22-25. Find which word in these verses is close in meaning to the word "mercy."

An old man was charged with a very serious crime. As the first session in the courtroom got under way, the young lawyer who was defending the man noticed that his client was very tense and nervous. Trying to comfort him, the lawyer said, "Charlie, you don't need to be afraid. I'm going to see that you get justice in the court today!"

The accused man replied, "Sir, it isn't justice that I want in this court—it's mercy!"

God is the great giver of mercy. He gives us what we need rather that what we deserve, because of the death of His Son, Jesus. That is mercy. Lamentations says that His mercy is new and fresh every morning. Thank Him for being patient with you and showing you mercy.

AUGUST 7   THAT WAS THEN, THIS IS NOW

On May 15, 1930, the first airline stewardesses boarded planes with the following set of instructions:
* Keep the clock and altimeter wound up.
* Carry a railroad timetable in case the plane is grounded.
* Warn passengers against throwing their cigars and cigarettes out the windows.
* Keep an eye on passengers when they go to the lavatory to be sure they don't mistakenly go out the emergency exit.

Of course, now things have changed, but the mission of the stewardess is still the same; take care of the passenger. Mark 16:15 tells us to *"go into all the world and preach the gospel to every creature."* Although times have changed, our mission remains the same, a mission we must strive to fulfill every day of our lives.

AUGUST 8   DRINK RESPONSIBLY—NOT!

Read Proverbs 23:29-35.

There is a popular phrase today to help curb drinking and driving, "Drink Responsibly." This implies that if you are going to drink, designate a driver who will drink in moderation so that he can drive everyone else home safely. Saying "drink responsibly" is like telling an elephant to walk lightly—it's impractical.

Drinking and driving is still the number one killer of young people in America each year. In 1993, more people died from alcohol related accidents on the streets of America than were killed in over 10 years of the Vietnam war. The Bible gives a clear warning to avoid intoxicating drink because of the negative implications.

Ask God if drinking helps or hinders your witness for Him and your relationship with Him. That's responsibility!

AUGUST 9   TO RUN LIKE A CHAMPION

Hupernikomen. Don't panic, it's not a Japanese cuss word. This is the Greek word for "conquerors" found in Romans 8:37 that says, *"...in all these things we are more than conquerors (hupernikomen) through him who loved us."* It means Olympic winners, overcomers. Like Orenthal. His name was the least of his problems. He was deformed because of rickets, a disease related to malnutrition. He was weak, soft-boned, and bow-legged. His head was disproportionately large. He involved himself in street gangs at age 13, but somehow managed to conquer the odds to eventually inscribe his name in the record books of the NFL. Orenthal James Simpson is best known as O.J.

What are your weaknesses? Your handicaps? With faith and perseverance, you can conquer adversity and be known as a champion for Christ!

AUGUST 10   THE MONEY PIT

Read Ecclesiastes 5:10.

You couldn't convince these guys that being rich is "where it's at." There is a lot of pressure today to be rich, even at a very young age. There is pressure to have nice clothes, drive a sharp-looking car and pick a job for the future that will leave you swimming in money. It's a trap!

It is interesting how natives of Southeast Asia used to catch monkeys to sell. They would drill 2 or 3 holes in a coconut, emptying out all of the contents. They would make sure the holes were just big enough for a monkey to put his hand through, but not big enough for him to pull it out if he had a clenched fist. Then, they would put sweet-smelling pieces of fruit inside and set it in a clearing on the edge of the forest. The natives would wait in a nearby hiding place until an inquisitive monkey had his hand in the coconut, grasping one of the pieces of fruit. They would then rush at the monkey, who could not get free from the heavy coconut (except by releasing the fruit) and capture him. All the monkey had to do was let go, but greed usually led to his downfall.

If you are not careful, greed will be your downfall as well. It will eat you alive like a cancer. Greed is a trap that will drag you deeper and deeper into its grasp.

Do you constantly think about how much you're getting paid or will get paid in the future? Are you constantly wanting more and more clothes? Are you always dissatisfied with what you have? These are dangerous signs that you may want to watch out for.

AUGUST 11   IN EXCHANGE FOR A CROSS

A young man was at the end of his rope. Seeing no way out, he dropped to his knees in prayer. "Lord, I can't go on. My cross is too heavy to bear." The Lord replied, "My son, if you can't bear its weight, just put your cross in this room. Then look around and pick out any cross you wish."

The young man was so relieved. "Thank you Lord," he sighed. Upon looking around the room, he saw many crosses, some so large their tops were not visible. Then he spotted a tiny cross leaning against the wall. "I'd like that cross, Lord!" he said while pointing it our. And the Lord replied, "But son, that is the cross you just brought in!"

Jesus said in Luke 9:23, *"If anyone would come after me, he must deny himself and take up his cross daily and follow me."* All of us have burdens. Sometimes they feel pretty heavy, but as we look around, we are made aware that others often carry much heavier burdens.

AUGUST 12   GOD IS THERE

> Remember me, God? I come every day
> Just to talk with You, Lord, and to learn how to pray.
> You make me feel welcome, You reach out Your hand.
> I need never explain, for You understand.
> I come to You frightened, and burdened with care,
> So lonely and lost and so filled with despair,
> And suddenly, Lord, I'm no longer afraid,
> My burden is lighter and the dark shadows fade.
> Oh, God, what a comfort to know You care,
> And to know when I seek You,
> You will always be there!

Read Psalm 55:16-18. Do you ever feel like David in the Psalm—that someone is waging a battle against you? In those times, you can rejoice just as David did, because God hears and rescues you. No one listens or understands as He does.

Is there some fear or burden that you need to give to Him? Do you just need to talk? He's waiting.

AUGUST 13   WHO'S THE BOSS?

Do you ever argue with God? As if we are smarter and know more than He! We can never go wrong if we will follow God's directions. He will never lead us in the wrong direction. He may lead us in a way we don't quite understand at the time, but it will always take us where we need to be.

One night at sea, the ship's captain saw what looked like the lights of another ship heading directly toward him. "Change your course 10 degrees south," he signaled. The reply came back, "Change your course 10 degrees north." This infuriated the captain, who signaled again, "I said change your course. I'm a captain and I'm on a battleship!" The reply came back, "And I said to change your course. I'm a seaman first class, and I'm in a lighthouse!"

God is our lighthouse and he said in *John 8:12*, *"...I am the light of the world. Whoever follows me will never walk in darkness, but will have the light of life."*

AUGUST 14   20/20 VISION

Read Isaiah 6:1-6.

Contact lens, glasses and binoculars all help people with poor vision to see clearly objects that are far away. Isaiah saw the Lord "high and lifted up" at a time of national disaster.

In 1963, John F. Kennedy was assassinated. America went into grieving! People all saw different things around his assassination; however, 40 years later, much remains unknown. Isaiah saw very clearly the Lord high and lifted up. In a time of political and social devastation, Isaiah looked up!

What do you see on your campus today? Drugs, rebellion, philosophical chaos, or do you see the Lord? No matter what financial, social, mental , physical or social problems surround you today, look up and see Jesus—He's always there.

AUGUST 15   HIDE AND SEEK

A popular youth speaker tells the story of playing hide and seek on a farm when he was younger. It was at night and everyone was scurrying about looking for that perfect spot to hide in while the person who was "it" counted. As time was about up, he found himself with no place to hide. Suddenly, he noticed a huge pile of hay, and he ran and dove head first into the hay to hide. He soon discovered that it was not hay, but a huge pile of manure instead.

It sure is easy to get in a hurry and "dive" into things before we realize what we are getting ourselves into. That is why Matthew 6:33 is so important. It says, *"But seek first his kingdom and his righteousness, and all these things will be given to you as well."* In other words, put God first and everything else, (career, mates, finances, etc.) will fall into place. Right priorities are important and necessary for success in the Christian life.

AUGUST 16   GOD KNOWS BEST

Why didn't God give her blue eyes? She had always wanted them, ever since she was a little girl. Why did she have to be stuck with ugly brown eyes? If it was up to her, she would change them, but it wasn't, thank goodness. She had another dream—to work with American Indians. What did that have to do with her eyes? Everything! When God did open the door for her to work with the Indians, as she had dreamed of for years, she learned a startling truth on her first day. In it she finally saw God's wisdom and truth about her eyes. The tribe she worked with had an ancient belief that was still held, which was all people with blue eyes were filled with evil spirits and must die immediately! That would have been her fate if she had had her wish of beautiful blue eyes, but God knew better. God in all His wisdom knew best.

Read Jeremiah 29:11. God always has your best in mind. He has made you as you are for a reason. He has allowed you to do or not do things for a reason. Ever heard the saying, "Father knows best?" Well, He (your heavenly Father) really does! Is there something about yourself you don't like? Trust God's wisdom in the way He has made you. Is there something going on that you don't understand? Trust God's direction.

The Book of James is full of wise, practical words that can help us on a daily basis. Look at what James says in 1:5, *"If any of you lacks wisdom, he should ask God, who gives generously to all..."*

A college professor handed back tests on which students had performed poorly. He began his reprimand, "A word to the wise..." One student who had really botched the exam spoke up and asked, "Is it all tight if the rest of us listen in too?" Hey, there is good news. If you have botched some of life's exams, you can simply ask God for wisdom, listen in, and receive the best advice known to man. Our daily prayer should be for Godly wisdom. It is free, and no one enjoys flunking.

Read Daniel 12:5-13.

On New Year's Eve every year in New York City's Time Square, the glass ball is lowered from a tower to signify the new year has come. At 5 seconds til midnight, the people will begin to countdown to midnight—5, 4, 3, 2, 1. "Happy New Year" is shouted, kisses exchanged and confetti thrown. The old has passed, the new has come! It's a wonderful time.

Daniel foresaw a time of ecstasy for those prepared for the end of time and a time of destruction for those unprepared. The end of the world has been discussed for centuries. From 1990-1993 our world has seen unprecedented events to point to the end as we know it to be—the fall of communism, the Persian Gulf War, the European Economic Commonwealth, The New World Order and many national disasters.

Theologically, it's 5 seconds to midnight! Are you prepared for the end? It will be far worse than anything ever known. To be ready, simply ask Jesus into your heart, repent of your sin and confess Jesus as Lord. 5, 4, 3, 2, 1...I hope you are ready!

AUGUST 19   SIGN OF THE TIMES

Have you ever paid attention to some of the road signs you see along the interstates and highways? "Slow Children at Play." Well, have them speed up, do some quickness drills with them. "Eat Here–Get Gas." Uh–Oh, better grab a can of Lysol. "Survey Crew Ahead." Excuse me sir, do you enjoy your job with the DOT? Or how about "Congestion Ahead." Well, time for some Dristan.

Fortunately, God has given us some clear road signs that mark the return of Christ. The disciples asked Jesus in Matthew 24:3, *"...what will be the sign of your coming and of the end of the age?"* Read those signs in the following verses. It is clear that we are living in the last days and have the hope of His soon coming to take us to heaven.

AUGUST 20   WHY IS MY LIFE SO BAD?

Read Habakkuk 1:2-6.

A question many people ask in crisis or adversity is, "Why do bad things happen to good people?" Habakkuk asked this same question. Where is the justice of God? I'm a good person! I'm not bad. Why do so many bad things happen to me? The correct response is, "What do good people do when bad things happen to them?" There are many ways to view a problem. The way you see yourself and your problems is the way you will respond in resolution. Is the glass half empty or half full? Is it a 20% chance of rain or an 80% chance of sunshine? Do not focus on the problem but on the Problem Solver, Habakkuk discovered in chapter 3, verse 1 that God would make a way of escape.

Whatever your storm may be, look to God and He will always look to you. Write down how you can give your problems over to Jesus!

AUGUST 21   THE LION'S ROAR

When you read Acts 4:29, you cannot help but think of a lion, the animal famous for its boldness, a characteristic which earned it the title "King of the Beasts." Kind of reminds you of some of the bold "Bad Boys" of the movies, like Clint Eastwood and his famous line, "Go ahead, make my day!" Or Rocky facing Apollo Creed,

Could you stand to be a little more bold for God? Our nation is in dire need of Christians who will take a bold stand for the statutes and directives of God's Word. So why not pray the prayer the Christians prayed here in Acts 4:29, *"Now, Lord, consider their threats and enable your servants to speak your word with great boldness*

AUGUST 22   ADAM & EVE WERE DRUGGIES

Read Genesis 3:1-7.

Did you know that Adam and Eve were drug users? You may say, "No way," but it's right in the Bible. In Genesis 3, the scripture says that Adam and Eve took a forbidden substance and put it into their body looking for a supernatural experience. They got what they went for—a supernatural reaction from substance use. They saw themselves as naked. However, instead of ecstasy, they were embarrassed and hid themselves from God.

Drug use in 1993 was on the upswing among colleges and universities. Drugs are used as a claw to pull people down the path of shame and addiction, from productivity to poverty. Drugs are a counterfeit to the real supernatural power surge that comes from a loving god. The scripture says in Acts 1:8, *"After the Holy Spirit comes on you, you shall receive power to be witnesses."*

The real power students long for is from above. Ask God how you can be induced for Him—the Most High.

AUGUST 23   ST. MATTRESS CATHEDRAL

Some experts say it takes 21 days to make something a habit. Others say it takes 5 weeks, still others say 6 weeks. What is so wild is that it only takes one day to get out of the habit of doing something.

Read Hebrews 10:25. It talks about the habit of church attendance, a touchy subject for college students. You go all your life and then all of a sudden, after starting college, you find yourself attending "Bedside Assembly" or "St. Mattress Cathedral." Sure it's tough finding a church like the one back home. Maybe no one else on your hall goes. Whatever the case, gathering together with other believers on a regular basis keeps us encouraged in the faith and strong in our Christian walk. *"Let us not give up meeting together, as some are in the habit of doing..."*

AUGUST 24   NO EXCUSE FOR DISOBEDIENCE

You know, it is always easy to make excuses for our sinful actions. Read about Saul's excuses in 1 Samuel 15:9-23.

Once, when the Chicago Bears were playing Buffalo, O. J. Simpson ran around the safety for a touchdown. The defensive coach was furious at his safety and shouted, "We pay you $75,000 a year to stop him!" The clever safety replied, "Well, they're paying him $400,000 a year to run around me."

We can think up a lot of excuses for blowing it in our Christian lives. We may say, "I'm emotionally tired right now," or "My parents don't treat me right," or "My friends always drag me down," but there are no good excuses for sin. Just as with Saul, the Lord is interested in our obedience, not our excuses. Write down one of your favorite excuses that you need to quit saying.

AUGUST 25   MAJOR LEAGUE FAITH

The greatest baseball player who ever lived never made it into the Hall of Fame. He never received any recognition and his name was never a household name, even though he got a hit every time he batted. He never missed a fly ball, never made a throwing error, never hit a double play, and never missed a grounder. How could such a great player not make it into the Hall of Fame? Because he never put his hot dog and drink down long enough to go from the stands to the field!

God has called us to be participants, not spectators. In James 2:17, it says *"...faith by itself, if it is not accompanied by action, is dead."* Talk is cheap. We must put feet to our faith!

AUGUST 26   THE NEED FOR INTEGRITY

The IRS once received a letter from a man who had cheated on his income tax returns. The man wrote, "I have not been able to sleep for several nights, so please find enclosed $200. If I still cannot sleep, I will send you the rest." His actions did not stem from a desire to be truthful and do what is right. He just did not want to feel bad anymore. Read Psalm 15:1-2 and write down one reason why we should desire to be truthful in life.

Psalm 15 says that we must be people of integrity and truthfulness if we want to have a dynamic, close relationship with the Lord. Is there any area of your life at this time in which you are not being entirely truthful? Are you tempted to cheat at work or at school? Are you tempted to lie for one reason or another? Do you have a lack of truthfulness affecting your relationship with the Lord? Write down an area in which you desire to grow in truthfulness.

AUGUST 27   AN APPEAL FOR ZEAL

A young boy was walking through the hallway of his church holding his father's hand. The hallway was decorated with several bulletin boards and Sunday School promotional materials. The item that caught the young boy's eye was a beautiful gold plaque. "Daddy what is that?" he asked. "Son," he began "that is a plaque dedicated to those who died in the service." The young boy pondered this for a moment and then asked, "Dad, was that the morning service or the evening service?" In 2 Corinthians 9:2, Paul is bragging on the church of Acaia and says, *"...your enthusiasm has stirred most of them to action."* Enthusiasm is contagious. It is something that we should spread everywhere. God has called us to be excited about the Christian life, not to be religious "deadheads". Stir someone today with your zeal for God.

AUGUST 28   HELP FOR TOUGH TIMES

Times are changing and getting a lot tougher. In 1940, the biggest problems teachers faced with their students were, talking, chewing gum, making noise, running in the halls, getting out of line and not putting paper in the waste basket. Sounds silly, doesn't it? The biggest problems with today's schools are completely differ-ent—drug abuse, alcohol abuse, pregnancy, suicide, rape, robbery and assault. Times have changed and you're going to school right in the middle of that change. It may seem overwhelming at times, just too tough to beat. It's not, though! Read Romans 8:31-37, 2 Kings 6:16 and 1 John 5:4. God says that you and He are a majority and that you can overcome anything in His strength. Times may be tough and the pressure may be growing , but it will never be too tough for God, and He's on your side!

Today, when things get tough, remind yourself that you and God can overcome the tough times. Keep repeating those verses in your mind. You are an overcomer!

AUGUST 29   NO FISHING!

Want to read something profound? Check our what Paul says in Philippians 3:13. He addresses a subject everyone can understand: the past. Too often, we allow Satan to remind us of all the mistakes we have made and to throw tremendous guilt up in our face. Paul states, *"...one thing I do; forgetting what is behind and straining toward what is ahead..."*

The famous preacher Dwight L. Moody once said, "God has cast our confessed sins into the depths of the sea, and He's even put a "no fishing" sign over the spot." How true! Let's press on.

AUGUST 30   OUCH, THAT HURT!

Read Ecclesiastes 10:1-4.

It's impossible to offend a dead man. You can scream insults and obscenities, bur he won't answer. He's dead.

Bill: "Since when did you stop singing in the choir?"

Charlie: "Since the Sunday I was absent, and everyone thought the organ had been tuned!"

If it's impossible for you to be offended, let me assure you, you will be offended. We must die to self to live for Jesus. After all, you can't offend a dead man! Are you dead? Write down how you can die to yourself today, to live for Jesus.

The young boys were fighting over a large bag of potato chips. Each boy wanted to hold the bag and ration out chips to the other. Their shouting match began to get physical and both boys grabbed the large bag of chips and started pulling. You guessed it. The bag tore, sending a shower of chips into the air and onto the ground. The two boys were each left holding half a bag with no chips.

So often we are more concerned about our rights and what we want, rather than thinking about others, that we rob ourselves and others of a blessing. Romans 15:2 gives us a great challenge for today. *"Each of us should please his neighbor for his good, to build him up."*

Read Hebrews 12:6. This scripture speaks of a subject that is not fun to discuss and often brings back painful memories. It talks of a father's discipline.

Can you imagine life with no traffic lights, no stop signs, and no speed limits? Your initial response might be, "Yeah, let's do it!" But no one would be safe. These boundaries are not intended to take away freedom or fun, but are designed to protect, just like God's boundaries found in His Word. *"He disciplines those He loves, and punishes everyone He accepts as a son."*

Several years ago there was an avalanche at Breckenridge in Colorado. Several skiers had ventured onto some unauthorized slopes which were out of bounds. They could have had fun on all of the slopes in bounds, but ignored the warnings and were tragically killed. Live within God's loving boundaries today and experience true freedom.

Read Genesis 11:1-9.

Have you ever had someone explain something to you and only get more confused?

What is believed to be the longest name in the English language is the name of a town in Anglesea, Wales. The name of this town is Llanfairpwelggwyngyllgogerychwrndrobwellhandyssiliogogoch. It has fifty-nine letters. In the postal directory only the first twenty letters are given. The name means, "The Church of St. Mary in a hollow of white hazel, near to the rapid whirlpool, and to St. Tisilio Church, near to a red cave."

The people who built the tower of Babel wanted to make a name for themselves, and pride entered their endeavor. Consequently, God confused their language. There will always be confusion and cloudiness when we try to live without God. Don't be confused. Jesus said, "I am the way, the truth and the life. No one comes to heaven, but by Me." That's clear!

In his rookie season as an NFL quarterback, Terry Bradshaw led the NFL in interceptions with 24 and was sacked 25 times for 242 yards. "They said I was stupid. And I believed them," says Bradshaw. But he credits adversity early in his pro career for his rise. Bradshaw went on to become the only quarterback to win four Super Bowls and was inducted into the Football Hall of Fame on August 5, 1989.

God has promised that we will be inducted into the halls of heaven if we will persevere in this life. Today's scripture says, *"For our light and momentary troubles are achieving for us and eternal glory that far outweighs them all."* 2 Corinthians 4:17.

Read Proverbs 29:20-24.

Have you ever been so angry that it provoked you to do something about it? We must channel our anger in the right way.

The woman who was instrumental in getting prayer removed from public schools wanted it banned also from outer space.

Mrs. Madalyn Murray O'Hair, after hearing words of prayer radioed by the three astronauts as they circled the moon, said, "I think that they were not only ill-advised but that it was a tragic situation..."

The noted atheist said she would register a protest with the National Aeronautics and Space Administration, which, she declared, had prompted the three test pilots in scheduling the prayer.

Unfortunately, the reason evil advances is that good men do nothing. Christians complain but do not act in mass. I want to encourage you to act out your anger against evil. Call your senator, representative or legislator and let them know how you feel! Your voice can and should be heard. If you don't like what the liberal media and entertainment industry are doing, register your complaint. Faith without works is dead. Do it now!

All of us, Christians or not, live each day under the law of sowing and reaping. Maybe horticulture is not your thing, but this law applies to every area of your life, especially in the area of finances.

An eye doctor stood in church to testify about his recent missions trip to China, a costly endeavor. He described various details of the trip and then explained why he would spend his own money in order to make such a trip. "Money is like manure," he explained. "You can pile it up somewhere and it will stink, or you can spread it around and do something for the Kingdom."

2 Corinthians 9:6 says, *"Remember this: Whoever sows sparingly will also reap sparingly, and whoever sow generously will also reap generously."*

SEPTEMBER 6   **ALL DRESSED UP & NO PLACE TO GO**

Read 1 Samuel 21:1-3.

You've probably heard the silly story of the little man who approached his doctor timidly and whispered, "Doctor, could you split my personality for me?"

"Split your personality? What on earth for?" the doctor asked.

The little man squirmed and said, "Oh, doctor, I'm so lonesome!"

Many times it is true that college campuses are filled with lonely students. Despite campus athletics, activities, various relationships and studies—it seems virtually impossible.

Be encouraged, strengthened and uplifted by the One Who is there for you to fill every void. Jesus is the lifter of your head. He is the strength of your life. You are never alone. Jesus is as close as the mention of His name!

A football coach walked into the locker room just before game time to find everyone suited up and ready to play except for his star quarterback. He was wearing tennis shorts and a T-shirt, along with tennis shoes and a baseball cap. The coach was shocked and furious. "Son, what are you doing?" the coach asked. "Well, Coach, everyone knows we are going to win this game today, so I decided to just play in these clothes."

Ridiculous, huh? That is about as ridiculous as a Christian facing Satan without the armor of God. Even though we know we will win, Ephesians 6:11 says, *"Put on the full armor of God so that you can take your stand against the devil's schemes."* Our war against Satan must not be taken casually.

Read Ecclesiastes 3:10-15.

Will Durant, on his seventieth birthday, says this, regarding the mixed blessing of a long life; "To live forever would be about the greatest curse imaginable!"

Mr. Durant was obviously not a born again Christian. As spirit filled believers, we can look at life the exact opposite of Will Durant. We have a "mixed blessing" and that is long life here, and eternity with Jesus, hereafter.

What will forever be like with Jesus? No pain, no disease, no politics—only Jesus! Praise the Lord!

Someone has said that grace is when we get what we don't deserve, such as salvation through Jesus. Ephesians 2:8 says, *"it is by grace you have been saved, through faith..."*

John is a good example of an act of grace. He seemed bent on rebellion and corruption. As a teenager, he joined the British navy but was soon arrested for desertion and demoted. He constantly mocked authority and associated with the lowest of the crew members. After sailing to Africa, John fell into the hands of a slave dealer and worked as a slave until escaping at age 21. He hopped aboard the Greyhound and continued his pattern of rebellion and wild living. One night, the Greyhound sailed into a furious storm. The ship was going down under the crashing waves, and for the first time ever, John prayed. He survived.

He retired from the sea to become a minister. Some time later, John Newton wrote these words:

> "Amazing grace! How sweet the sound,
> That saved a wretch like me!
> I once was lost, but now am found,
> Was blind, but now I see!"

SEPTEMBER 10   PRAY OR PAY

Read 2 Chronicles 7:14.

The Bible has come under severe attack in recent years, from arguments of authenticity to relevance. It sustains them all.

It is illegal to read the Bible in the public schools of Illinois, but a law requires the state to provide a Bible for every convict! Don't worry, kids. If you can't read the Bible in school, you'll be able to when you get in prison!

How sad but very true this is! Since 1963, when prayer and Bible reading were ruled unconstitutional by the Supreme Court, America has paid the price. Exponential increase has taken place in theft, assault, murder, disease, rape, and on it goes. Prayer and Bible reading have been proven to be effective.

Pray or pay—the facts are in! The last thing on a couple's mind when they leave church after a wedding is divorce. Without prayer and scripture reading, love can grow cold.

Pray or pay. Which way will you go?

Let's face it, most people desire power in some form or fashion. For some it may come in the form of an important job; for others, it may come in the form of having their own credit card, a place to live or nice car. Remember the old ads in the back of the comic books that promised a powerful physique within a few weeks? Or the classified ads that guarantee you will own your own business with a $50,000 a year salary in six months?

When you get down to it, the only power we should strive to possess is the power talked about in Acts 1:8. It says, *"but you will receive power when the Holy Spirit comes on you..."* Power to do what? Make money? Gain status? No, power to do the only thing that really counts, to be a witness for Christ and expand the Kingdom of God.

Sometimes the difficult tasks of life seem insurmountable. "It's impossible. There's no way. I can't do it."

Eugene Ormandy, music director of the Philadelphia Orchestra, commits musical scores to memory. He says he developed his powers of total recall as a child in Budapest.

His father was a dentist who was determined that his son should be a great violinist. "I hit on the idea of memorizing the music," explains Ormandy, "so that I could read novels as I practiced. It came easy and has been ever since."

Just as Ormandy has done, what we love and enjoy, we somehow make a way for. Ormandy was forced to practice his music, but loved to read. He made a way to accomplish what he really enjoyed through seemingly overwhelming circumstances.

What are you making a way for? Through difficult and overwhelming pressures of college life, you can overcome. You can be a sold-out, Bible believing, college student. You can accomplish God's purposes.

A teenager mustered up the courage to ask his grandfather a question about sex. "Granddad, your generation didn't have AIDS and all of these social diseases. What did you wear to have safe sex?" "A wedding ring," the grandfather answered.

Take time to read what Colossians 3:5 says about sexual immorality. The number of AIDS cases among teenagers is doubling every month. The rate of heterosexual transmission of HIV has increased 44% since 1989. 63% of all sexually transmitted diseases occur among persons less than 25 years of age. The "safe sex" solution is no solution at all. The Biblical sex solution is the only safe sex. Determine to not become a statistic or be a notch in someone's sexual belt. God has a better plan for you.

Read Leviticus 13:1-5.

As a college student, you face unique crossroads, one being the widespread sin of promiscuity.

Venereal disease is climbing by leaps everywhere. Cases of gonorrhea has topped the one million mark. About 100,000 women each year are made sterile by gonorrhea infections. The dean of the School of Public Health at UCLA considered it "out of control." At least one strain is completely resistant to treatment by penicillin. The main reason: increased sexual promiscuity and homosexuality.

These facts regarding VD are alarming but hopefully are not your reason for avoiding sexual promiscuity. You need to know the facts, but more than that, you need a firm foundation in Christ to keep you pure.

Commit again, even now, your strong stand to live righteously in every area of your life. You can do It!

SEPTEMBER 15   YOU CAN TAKE THAT TO THE BANK!

In 1989, 109.5 million people used 956.9 million credit cards to buy $430.3 billion worth of goods and fall $206.7 billion into debt. In 1990, America's 27.4 million teens spent $79 billion on products and services. One college student's philosophy was "Charge it and pray for the rapture!" Posted in dorm rooms across America is the saying, "When the going gets tough, the tough go shopping!"

Money is a touchy subject (especially when you don't have any), but one that Jesus addressed often. What kind of steward are you with the money God has given you? Jesus said in Luke 12:34, *"For where your treasure is, there your heart will be also."* In other words, if God really has a hold of our heart, then he has our wallet or checkbook, too. We must discipline ourselves to handle our finances in ways that will please the Father.

SEPTEMBER 16   THINK BEFORE YOU SPEAK

Read 1 Samuel 3:9-10.

Sticks and stones can break my bones, but words will never hurt me. How I wish this were true. Words can cause much pain.

It is said of Julius Caesar that, when provoked, he used to repeat the whole Roman alphabet before he allowed himself to speak.

This does not sound like a bad idea. Guard your tongue, at all costs, even if it means as Caesar did—repeating the alphabet when provoked. Whatever it takes to slow you down and think before speaking.

The tongue is the most unruly part of the body the Bible says. Think, if you can conquer your tongue, what an advantage you will have! How can you overcome the urge to say the wrong thing?

We live in a day and age where commitment seems to be a thing of the past in so many ways. Many young people have a difficult time making solid, long-term commitments because they have not had many examples of those who do.

However, God's mandate and, according to Christ the greatest commandment, is reiterated in Mark 12:30. It says, *"Love the Lord your God with all your heart and with all your soul and with all your mind and with all your strength."* There is no place for half-hearted commitment in the Christian life.

A pig and a chicken saw a sign at a local restaurant that read, "Bacon and eggs–All U Can Eat!" The chicken said, "What do you say we help the cause?" "No way!" said the pig. "For you that is just a donation, for me it means total sacrifice!" Are you willing to make a total sacrifice for God?

SEPTEMBER 18   *IT IS A NEW DAY*

Read Exodus 31:12-18.

As a college student you may have experienced loneliness. Know today that you always have a friend in Jesus. He is One Who sticks closer than a brother. It does not matter if you hear Him speaking to you or not, He is there!

In a cellar in Collogne, Germany, after World War II, were found these words on the wall:

I BELIEVE...

I believe in the sun,
   even when it is not shining;
I believe in love,
   even when I feel it not;
I believe in God,
   even when He is silent.

A young high school graduate became restless with everyday life on the family's farm in Kansas. He finally worked up his nerve and approached his dad, a hard-working, God-fearing man. "Dad, I know you've always talked about the day that I will be running the farm. But I've made up my mind to go to Chicago to find a better job." He cashed in the CD his father had for him and headed for "greener pastures." The $9,000 dwindled to only $50 within four months of "living it up," and he soon found himself with no rent money and barely enough food.

Luke 15 tells us a similar story of a young man who took his inheritance and blew it in the big city. He found himself eating pigs' food. Verse 17 says, *"...when he came to his senses..."* Sadly enough, many people have to reach bottom before realizing their need for a personal relationship with Jesus. The prodigal son returned home to his father, and his father received him with open arms. That is exactly what our Heavenly Father does when we repent and return to Him.

SEPTEMBER 20   WINNING, WOUNDED OR WIPED OUT

Read 1 Kings 2:1-10.

A king, a throne and a dying monarch. David had led Israel for years and, before his death, he had a coronation for Solomon. Words of advice and wisdom from a falling work horse to a young stallion were given. These words were:

1. Act like a man of God.
2. Be true to the Word of God.
3. Rely of the promises of God.
4. Execute the judgements of God.

As you go day to day, this advice is still worthy of application. Write down how you can live these principles on your campus today.

It is interesting to read about explorers in the past who were in search of a fountain of youth. The search continues, but instead of searching for a fountain somewhere in the wilderness, we look in the cosmetics department or at the fitness club. Being young can have its advantages as well as its disadvantages.

People often say that today's youth are tomorrow's leaders, but Paul seemed to believe that youth could be the example. He encouraged Timothy, his young disciple, in 1 Timothy 4:12 by saying, *"Don't let anyone look down on you because you are young, but set an example for the believers in speech, in life, in love, in faith and in purity."* There has never been a greater opportunity for today's youth to go against the tide of immorality and set the example as today's moral and spiritual leaders.

Read Genesis 3:10-15.

Have you ever noticed that things are not always as they appear?

From Turin, Italy came this news: Ernesto Caraldi, age 77, had been found apparently lifeless by neighbors. After a doctor pronounced a death certificate, the undertakers put him in the coffin. When he suddenly sat up and asked why the undertakers had been called, he collapsed and died of fright.

In your daily campus routine, try not to get disillusioned or frightened by what you see and hear. Trust Jesus and He will show you the way of truth!

"How could God ever forgive me?" cried a young college girl. "You can't imagine the awful things I have done!" God's grace and forgiveness is indeed a mystery, but even if we don't understand how

God can wash away our sins, it does not change the fact that He does if we accept His forgiveness by faith. In 2 Corinthians 5:17 it says, *"If anyone is in Christ, he is a new creation; the old has gone, the new has come!"*

A small town in Central Florida decided to make something useful from garbage. They turned the landfill where all the garbage was dumped into an 18-hole golf course, complete with beautiful fairways and landscaping. There is no sign of the old garbage and visitors to the community never knew what was once there. God covers our garbage and makes something beautiful of our lives.

SEPTEMBER 24   BUDGET DELUXE MOTEL

Read Numbers 6:1-8.

When you hear the phrase, "Budget Deluxe Motel," does, something not seem right? Is it budget or deluxe? Can you have both? There's a similar question about life. Can you love and serve Jesus and yet continue living in a worldly fashion?

The scripture is clear that there are many things that we should be separate from in life. We cannot have it both ways, either we are in or we're out. Ask yourself right now, "Am I trying to have it both ways?" Jesus came to give us life, but not the life of the unrighteous.

SEPTEMBER 25   LOVE IS A CHOICE

The event made the headlines and top billing on the nightly news. Two police officers had been suspended for pouring oil on a homeless man and roughing him up a bit. The community responded with anger and embarrassment over the senseless "hate act."

"Dear friends, since God so loved us, we also ought to love one another." These are the words found in 1 John 5:11. So many people are under the impression that love is a feeling. Not so. Love is a

choice, and when we choose to love, it makes us feel good. We are to follow God's example. He chose to love us, and we are to choose to love others each day, regardless of the outward differences we may see.

SEPTEMBER 26   THE FOOTPRINTS OF THE CROWD

Read Ezekial 7:10-14.

At the Riverfront Coliseum in Cincinnati, Ohio, "The Who" came to put on a concert. The crowd, anxious for the doors to open, began to get restless. Someone shouted from the back of the line, "The doors are open!" All of the crowd began to push and the doors gave way to the crowd. People were being knocked down, stepped on and trampled. Eleven people died that night because of the footprints of the crowd.

Acceptance is a god in America. If I need to change for approval, I may do that. I may change my language, my apparel and my actions to gain acceptance. As fast as the crowd approves, it can also trample. Are you wearing the footprints of the crowd? Are your peers directing you or is the power of God in control? Can you be an influence on others for Jesus, instead of allowing them to influence you?

SEPTEMBER 27   THE DATING DILEMMA

Christy angrily hung up the phone after the conversation with her dad turned into a heated discussion over Ryan. Christy was a sophomore at the University and felt she was old enough to make her own decisions about dating. Ryan was a junior and, unlike Christy, had not been raised in a Christian home. Christy's dad did not accept her "missionary" concept of dating, even though she tried to convince him that everything would be okay and she would be good for Ryan.

She had often heard her youth pastor quote 2 Corinthians 6:14. *"Do not be yoked together with unbelievers."* The words echoed in her head, but she was strong and well grounded. Within months however, Christy had stopped attending church, had become sexually active with Ryan, and was attending parties she once deemed "off limits."

This scripture asks an important question, *"What fellowship can light have with darkness?"* We cannot compromise in this vital area of our lives.

SEPTEMBER 28   LOOK WHOSE HERE!

Have you ever been surprised and heard someone say, "Look whose here"? Read Joshua 5:13-15.

I remember taking my first drink of beer as a 15 year old boy. One drink followed another until I was totally inebriated. Having "fun" at a party, I heard a friend say, "Look who's here!" I remember the humiliation and embarrassment. I learned my lesson.

Very soon Jesus will come again and the whole earth will say, "Look who's here!" No matter what Popular Science, professors, or atheist say, He's coming soon! How can you prepare for His coming?

SEPTEMBER 29   THE ART OF PRESERVATION

It is natural to want to live, to remain alive, to preserve life. Efforts are made daily in the medical field to extend life. However, a different law is present in spiritual matters. Luke 17:33 says, *"Whoever tries to keep his life will lose it, and whoever loses his life will preserve it."*

Just off I-91 in Connecticut is Dinosaur State Park, where over 200 dinosaur footprints, believed to be 185 million years old, are preserved in the sandstone. You can even take home a plaster-of-paris cast of a footprint. The scripture is clear, though, that if we try to hang on to our lives, it will slip right through our fingers. The art of preservation for us is total surrender. If we give our lives totally to the Lord, we will live forever with Him.

SEPTEMBER 30   STARTING ALL OVER AGAIN

There are few things more exciting than a new baby being born into the world. The miracle of birth has awed mankind for centuries. John 3:3 paints a word picture of the birth experience in a conversation Jesus is having with a man named Nicodemus. Jesus said, *"...no one can see the Kingdom of God unless he is born again."* Nicodemus then asks, "How can a man be born when he is old?"

I doubt that many of us would want to relive the experience of entering this world stark naked via a cold hospital room with bright lights and strange faces, but Jesus wanted Nicodemus to understand the newness of life as it relates to salvation, and that spiritually we must experience new birth. Now that is exciting! According to Jesus, being born again insures us a place in heaven.

Read Revelations 2:4-5. You talk about getting down to the real truth, the bottom line—here it is! Do you remember the first time you fell in love? A husband and wife were riding down the road and the wife said, "Honey, do you remember when we first fell in love and we would sit real close? You would put your arm around me while you were driving and I would snuggle up to you?" The husband smiled and said, "Yes dear, I remember." "Why don't we do that now?" the wife asked. "Well, dear," he replied, "in 20 years of marriage, I've never moved!"

John writes, *"You have forsaken your first love. Remember the height from which you have fallen! Repent and do the things you did at first."* At times, we feel far from God; yet God has not moved. We simply need to do the things we did when we first began following Him and return to our first love.

Read 2 Samuel 11:2.

Models, magazines, and movies say that beautiful, thin and blemish free is the way to be—making the outside beautiful.

A man who had traveled in Europe says that while many of the old cathedrals in France are beautiful on the inside, their exterior is often distasteful because of the art carvings which depict animals with hideous features. Inquiring about this strange custom, he was told that the builders in the Middle Ages wanted these figures to represent man's carnal appetites and prejudices! They were placed there to remind all who came to worship that they should leave bitterness and wrath outside the sanctuary if they hoped to receive God's blessings.

If young adults are professionals at anything, it is in wearing a mask. Painting a smile on the outside when things are unraveling on the inside. When Jesus comes inside, He changes us from the inside out, not just the outward appearance. True beauty can be accomplished when we allow His beauty to work within. Look in the mirror. You're the only you God made. Be the best you can be!

OCTOBER 3   LIVING ON THE EDGE

Playing with sin is like playing with fire. You will eventually get burned. Many people think it is exciting to play "footsies" with sin, but not get involved. They want to see just how close they can get without giving in to temptation. The Bible tells us in 2 Timothy 2:22 to *"flee the evil desired of youth..."* We are to run from evil.

Three men were applying for a job as a truck driver. The employer concluded the interview with the question, "If you were driving a truck filled with explosives on a narrow mountain road, how close to the edge could you drive and still deliver the load?" The first applicant bragged of his ability to drive within a foot of the edge. The second driver stated he could hang the tire over the edge. The third said, "Sir, I respect the danger and would stay as far from the edge as possible!" He got the job.

OCTOBER 4   PEARLS BEFORE PIGS

Read Proverbs 19:22.
93% of Americans profess to believe in God. 47% of Americans attend church. 27% believe in a real devil. The Bible is filled with facts and roadways for success. However, many trample underfoot those truths.

Professor George Lyman Kitteredge of Harvard, the Shakespearean scholar, was once annoyed by the students noisily preparing to leave the class the moment the bell sounded.

"Just a minute, gentlemen," he said, "I have a few more pearls to cast."

Do you hear the truth but ignore its application? Some come 12 inches short of God. They have it in their head but not 12 inches down to their heart. Are you hearing and obeying God's truth? Write down how you can apply God's truth to your life today.

OCTOBER 5   HURRY UP AND WAIT

You have probably heard about the girl who was anxious about meeting her mate for life. She was a college senior and had grown impatient waiting for Mr. Right. "God," she prayed, "give me patience and give it to me NOW!" It was for her and others like her that Hebrews 10:36 was written. It says, *"You need to persevere so that when you have done the will of God you will receive what he has promised."*

Let's face it, we live in a fast world, one that encourages impatience. However, God is not into "microwave" Christianity. He wants us to learn to wait on His timing and for His perfect plan to unfold in our lives.

OCTOBER 6   YES, YOUR HONOR

Read Numbers 27:20.

Have you ever been told to do something you didn't want to do? There is something distasteful to have others demand we do things by their agenda.

Judge John A. Weeks spotted a man sitting in the rear of his Minneapolis courtroom wearing a hat. Disturbed by this disregard for courtroom decorum, he ordered the man to leave.

Then the clerk called for the burglary case of George A. Rogde, who had been freed on bond. Rogde didn't come forward.

"Your honor," said the prosecuting attorney, "that is the man you ordered from the courtroom."

Police are still looking for Rogde.

The Judge spoke, the man obeyed and he was set free. Everyday God is trying to tell us which way to go, which things to do and if we do, freedom will come. How can you listen to God's instructions—obey and be free!

OCTOBER 7   PRICELESS TREASURE

In the late 1800's, the United States was presented with a valuable, priceless gift from the country of France known as the Statue of Liberty. It is admired by Americans year round and highly valued by our nation. Can you imagine the insult it would have been had America declined this gift? "Ah, no thanks. It is nice, but we just don't have anywhere to put it."

Years ago, God thought of the very best gift He could give to mankind. John 3:16 has been taught in many a Sunday School class. It says, *"For God so loved the world that he gave his one and only Son, that whoever believes in him shall not perish but have eternal life."* God gave the most priceless, valuable gift He could give, because He loves us. Can you imagine the insult when thousands reject this treasure day after day? Let us, today, receive His gift and realize the value of Jesus in our lives.

OCTOBER 8   IT'S IN THE AIR

Read Exodus 15:25-27.

There are millions of germs and infections that fly through the air at all times. The flu, hepatitis, spinal meningiti,s and other infections can be passed by body contact or simply by breathing in the germs.

Fifteen nations have chemical and biological warfare programs. Major General Marshall Stubbs estimated that an aircraft carrying

10,000 pounds of germs could kill 60 million people. Russia and the U.S. have together enough biological weapons to kill every living thing on earth.

We are subject to our environment. It has been said what we associate with, we become. Many doctors contact the disease that they treat. Sports fans are elevated to frenzy levels by sitting in the stands. Christians can be influenced by their surroundings too. Are you being influenced or are you influencing others? Ask God how you can contaminate someone else with His love for them.

OCTOBER 9   THE BEST POLICY

The elderly sixth-grade teacher spoke the words almost daily, "Oh what a tangled web we weave, when first we practice to deceive." Yet by the time the students reached college, the words were long forgotten or ignored. A recent study revealed that a third of college students admit to cheating on an exam within the past year. More than a third of all high school and college students said they would lie on a resume or job application to get a job. 21% of college students indicated they would falsify a report if necessary to keep a job.

Ephesians 4:25 says, "*Therefore each of you must put off falsehood and speak truthfully...*" The practice of deception spins a web that will ensnare and entangle a person for a lifetime. We must put off or throw off lying and be people of integrity and truth.

OCTOBER 10   PUT IT ON MY TOMBSTONE

Death is a part of life. What would you like to have on your tombstone?

The most alcoholic person was in England. A person by the name of Vanhorn averaged over four bottles of ruby port daily for 23 years prior to his death at 61. He emptied 35,688 bottles in his lifetime. His tombstone read, "Born 1903. Died 1964. Held more drink than anyone before."

What a sad accomplishment for life. He could drink more than anyone? What will yours say, "Born, lived, and did what?" How do you want to be remembered?

OCTOBER 11   FAITHFUL OR FAMOUS?

What do you think of when you hear the word "famous"? Madonna? Michael Jordan? The Mona Lisa? Niagara Falls? Everyone dreams of being famous, but God would much rather us be faithful if it comes down to the two. You have probably heard of the Seven Wonders of the World. What is so interesting about these seven wonders is that only one of them had any real use. Famous though they be, they were not much good for anything.

Paul says in 2 Timothy 2:21, *"If a man cleanses himself, he will be an instrument for noble purposes, made holy, useful to the Master..."* Wow! Forget being famous. Can you imagine a person who is described as being useful to the Master? Fame is sweet, but being good for something, an instrument God can use, should be our goal.

OCTOBER 12   THESE ARE THE 90's

Today, liberals, moderates, and conservatives are constantly trying to set the standard of what is and what is not correct in society.

A few years ago, a group of people in Missouri founded a town and named it Liberal. They were so extremely "liberal" that churches were not to be allowed. In their boom literature they boasted that it was "the only town of its size in the nation that did not have a church." The town died 5 years later.

America was founded on religious and political freedoms and the church was significant in the direction of a sprawling new nation.

Today, the church is under attack like never before. Religious freedoms are removed, attacked, and ridiculed. If we do not stand for the church in the 90's, there will be fewer freedoms for our children. Don't let your town be named "liberal." Write down how you can influence your city.

OCTOBER 13   AMONG THE TOMBS

Has anyone ever asked you a question that just reaches out and slaps you like a cold wind? Read Luke 24:5 and you will discover just such a question. Some women had walked to the tomb where Jesus had been laid (girls always travel in clusters) to prepare Jesus' body for burial. Two angels asked, *"Why do you seek for the living among the dead?"*

So often, we are guilty of the same thing. We seek for life among dead things. There is no life in crack cocaine, only death. There is no life in a can of beer, or in immoral sex, in money, relationships, or fame. We must get out from among the tombs and realize that life is found in Jesus. He is the creator of life, the giver of life eternal.

OCTOBER 14   POLITICAL CORRECTNESS—NOT

In the 90's a catch phrase has become "politically correct." This means things should be gender and circumstantially correct for all situations.

The official state seal of Georgia shows three pillars supporting an arch of the Constitution, which are labeled "Wisdom," "Justice," and "Moderation."

By mistake, the Crown Cork and Seal Company of Baltimore, which prints and furnishes the beer bottle caps to bottlers, printed the motto: "Wisdom," "Justice," and "Salvation." For several years this error went unnoticed, then it was discovered by a reporter of the *Evening Press* and the printers assured the Malt Beverage Unit of the

State Revenue Department that they would immediately correct the error.

The Gospel is not politically correct. Steven was stoned for his presentation, and Jesus was crucified. Are you reeling from political correctness or are you presenting God's Word pure and clear? Take a stand. If we don't stand for God's Word, we will fall for everything. How can you be biblically correct today?

OCTOBER 15   *A DEADLY DISEASE*

Did you know that if you took half a million viruses and laid them end to end, they would make a line no longer than the word "virus" itself? Three thousand million billion of them would weigh an ounce, but only one small virus can bring so much misery. They can even destroy a person. Sin is the same way.

So often, we view sin as such a little thing. We think that it will never "infect" us or cause us to be spiritually diseased. In James 1:15, it says, *"Then, after desire has conceived, it gives birth to sin; and sin, when it is full grown, gives birth to death."* Just like a deadly virus, sin grows into a killing disease, physically and/or spiritually. We must develop an immunity, a hatred for sin, and walk in righteousness and holiness.

OCTOBER 16   *LET'S GO TO DISNEY WORLD*

Read Psalm 119:9-11.

After winning a national or world championship, athletes are asked, "Where will you go after the game?" A popular response is, "Disney World!"

The University of North Carolina broke a 130 year tradition on graduation day in 1972 by not handing out Bibles to graduating seniors along with their diplomas. The institution's trustees' executive committee voted to abolish the practice, which had been questioned on constitutional grounds. The Bibles had been handed out with the diplomas at Chapel Hill since 1842 and had been purchased through student fees.

If we were handing trips to Disney World, instead of a Bible, there would be no questioning of its intent. However, when you give God's life-giving Word to a person, it shakes the corridors of hell. Carry your Bible with you and make it visible for others to see. I promise, just as in North Carolina, it will provoke a response. Do it today!

OCTOBER 17   KINDNESS REPAID

It pays to be kind. In a cruel, often cold world, kindness shines like the sun. A businessman pulled into a motel on a country highway after a long day of traveling. The desk clerk informed him that all rooms were full, but if he would like, he could sleep on the extra bed in his small room. The tired businessman accepted, and before leaving the next morning, thanked the man and told him if he ever owned a hotel, he wanted him to run it.

Years later, the desk clerk got a call from the businessman. He had bought a hotel in New York City and, as promised, wanted to hire the clerk. The businessman was John Jacob Astor. The hotel was the Waldorf Astoria.

Read Ephesians 4:32. *"Be kind and compassionate to one another, forgiving each other, just as in Christ God forgave you."*

OCTOBER 18   LET'S CLEAR THE AIR

Anxiety, depression, and unresolved anger can lead to emotional and physical problems. We must find a way to vent our emotions in a constructive way.

A college professor answered his telephone at 3:00 am. "This is your neighbor, Mr. Smith," said the voice. "Your dog is barking and keeping me awake." The professor thanked him kindly and hung up. The next morning Mr. Smith's telephone rang at exactly 3 am. "This is the professor," said the caller. "I just wanted you to know that I don't have a dog!"

There are appropriate ways of expressing our emotions, others can be devastating. Are you carrying around unwanted emotional baggage from yesterday's failures and disappointments? You can release your anger and fear to Jesus today.

OCTOBER 19   A SECRET TO SUCCESS

One of the growth opportunities of being a college student is learning to make it in sometimes less-than-luxurious conditions and circumstances. Some dorm rooms equal that of a nice walk-in closet, a budget is something that is only talked about in Accounting 101, and steak and lobster dinners are something you see in a magazine ad. In Phillipians 4:12, Paul states, *"...I have learned the secret of being content in any and every situation..."* He leaves no room for complaining or groaning about how bad things may be.

One lady wanted a shade tree and called a local nurseryman for help. "I want something that will not spread too much but will give shade, will not shed leaves everywhere, and will allow me to see the sun in the winter." The nurseryman shook his head and said, "Lady, you don't want a tree, you want an umbrella!" We can learn to be content with whatever we have!

OCTOBER 20   THE GOLDEN CALF

Most American students would say that idolatry is a third world occurrence. However, It is alive and strong in our own back yard.

A visiting minister was substituting for the famed pastor, Henry Ward Beecher. A large audience had assembled to hear the popular pastor. At the appointed hour, the visiting minister entered the pulpit. Learning that Beecher was not to preach, several began to move toward the doors. The visiting minister stood and called out, "All who have come here today to worship Henry Ward Beecher may now withdraw from the church! All who have come to worship God, keep your seats!" No one left.

Do you give your devotion to the Master of His messengers? If we are not careful, we can misappropriate our worship for God to inappropriate objects. Take a moment right now and commit to worship Jesus and Him only.

OCTOBER 21   PASS THE SALT, PLEASE

In Matthew 5:13, Jesus pays a high compliment to those who belong to Him. He says, *"You are the salt of the earth..."* That may not be your idea of a compliment and certainly not the most charming thing to say to your boyfriend or girlfriend, but can you imagine life without salt? Americans consume thousands of pounds of salt each year. French fries, scrambled eggs, grits, or green beans just wouldn't be the same.

Salt does four things. It flavors, heals, preserves, and purifies. So Jesus is saying that we are the agent, the additive that brings flavor, healing, preservation of righteousness, and purification to this world. We can turn a tasteless world into one that is tasteful. What a compliment!

OCTOBER 22   IT'S BETTER TO GIVE THAN TO RECEIVE

The 90's are a "me first" generation. Contributions to charities are at all time lows. Looking out for number one can cost in the long run.

The manager of large department store in the Middle West told of a middle-aged gentleman who spent a great deal of time and money in his toy department in the days before Christmas. He watched for poorly dressed children. When he saw them looking longingly at some gift, he wrapped up the toy and presented it to the child.

He had no children of his own, but he made many little ones happy, who might never have received a single present.

Are you a giver? It's much more fun to give than to receive. Try it today. Look around you and see someone in need, then give. Psychologists say it's a key to emotional stability. Give and it will be given to you!

October 23   *Armed and Dangerous*

When the United States invaded Iraq and fought in the Persian Gulf January of 1991, the U.S. Military forces used many weapons in order to defeat their foe. One of the most remembered was the Patriot Missile, launched to intercept enemy missiles in mid-air.

As Christians, we fight in a war against an evil enemy every day, and the Bible makes it clear that we also have weapons. God has equipped us for battle, and we are destined to win. In 2 Corinthians 10:4 it states, *"The weapons we fight with are no the weapons of the world. On the contrary, they have divine power to demolish strongholds."* We don't fight with guns, tanks, missiles, knives or grenades. The weapons God has given us, the Word, prayer, praise and the blood of Jesus, are much better. They have divine power. The choice we must make is not whether we will go into battle. We are there daily. It is whether we will use the weapons God has given us.

Do you ever feel the need to get away from your immediate surroundings to experience something new?

Sinclair Lewis told about the time, when crossing the Atlantic, that he saw an old lady on deck reading his latest book about which there had been some hot discussion. By the number of pages she had read, he judged that she was approaching the shocking passage which had caused the most trouble, and he kept an eye on her, to see how it would affect her. Presently the old lady rose up, walked firmly to the rail and flung the book far into the ocean.

Statistics prove that we will be what we associate with. Our closest friends, acquaintances, or our surroundings can cause us to accommodate rather than dictate our response. Do you need to get away from some negative influence in your life today? Nike says, "Just do it!" Do it now and don't delay!

Read 1 Timothy 4:1 and you will get a fairly accurate barometer of our day and time. It reads, *"The Spirit says that in later times some will abandon the faith and follow deceiving spirits and things taught by demons."* Have you known someone in recent months or years who has done just that? It is alarming the number of people who are walking away from their beliefs and being deceived by teachings that are blatantly contrary to God's Word.

The New Age religion is a strong lure to people who want to "find themselves" or be "enlightened" to the truth. There is nothing new about the New Age movement. It is the same form of deception that Satan used in the garden with Adam and Eve, that there are no absolutes and that everyone possesses or can possess deity within self. Stars like Shirley Maclain and John Denver, psychic hot-lines, and New Age stores have given rise to its popularity. Remember the old saying, "If I'm a fool for Christ, who's fool are you?"

Adam and Eve's mistake of eating the forbidden fruit has caused much death and decay.

The first Europeans arriving in America introduced the Indians to liquor. The Indians gave the white men tobacco, which they had used sparingly before.

The Indians developed a fondness for "firewater" which persists to this day. There is a high incidence of liver ailments among them. Whites developed tobacco into a great industry, and now, lung cancer, virtually unknown among Indians, takes thousands of lives among white smokers.

Today, we feel the implications of the original sin of Adam and Eve. Not all things are good for you. Just because God created it, that doesn't mean it should be put into your body. Are you allowing yourself to inject improper products? An apple a day can kill you! Protect yourself by only partaking of that which pleases God.

What chances could a slow, short, white kid from Oklahoma have of ever making it in big, college basketball, much less the NBA? Many people believed none, but Mark Price of the Cleveland Cavaliers was determined to make believers out of them and beat the odds. Even after a successful career at Georgia Tech, Price was still bypassed completely in the first round of the NBA draft. Now, as the starting point guard for the Cavaliers, Price has an NBA career average of 15.9 points per game and 7 assists. He also has the second highest single season free throw percentage in NBA history.

What is the key to beating the odds and having a successful walk with God? Are there people or circumstances that say you cannot make it? 1 John 5:4 says, *"...This is the victory that has overcome the world, even our faith."* Faith in God and in His promises will insure victory in our lives.

October 28   You Can't Judge a Book by its Cover

Bookstores across America display all kinds of books and magazines in various topics. However, no area has grown more rapidly than books surrounding the topic of occultism.

Occult book sales doubled in the last four years. The occult movement has its own trade magazine called Occult Trade Journal.

One of the busiest bookstores is Metaphysical Center in San Francisco, selling about $12,000 a month. It offers courses in palmistry, reincarnation, astral projection, numerology, and others. The film *Exorcist* gained $70 million in a year, and the book by that title has 10,000,000 in print. Big colleges and universities offer credit courses on the occult.

Everyone wants to know about the supernatural. Is there life after death? Is there a real heaven and hell? Jesus talked about hell twice as much as He talked about heaven. People who say they don't believe in hell won't make it go away. Everyone wants to believe that they are going to heaven. There's only one way to get there, have faith in Jesus, obey His commands and avoid the appearances and actions of evil. You can judge a book by its cover. If it looks evil, sounds evil, it probably is evil. How can you avoid evil today?

October 29   A No–Risk Investment

A local bank had removed any tattered or worn currency from circulation and placed it in a bag to return to the U.S. Treasury. A $1 bill and a $20 bill began conversing about all they had done and seen. "I saw fourteen different countries, nice restaurants, several fine department stores, and the country club here in town," said the $20 bill. "Where have you been?" "The only place I ever went was church," replied the $1 bill.

Statistics show that the number of people and the amount given to churches has decreased in the last year. 2 Corinthians 9:7 says, *"Each man should give what he has decided in his heart to give, not reluctantly or under compulsion, for God loves a cheerful giver.* " Maybe misuse of church funds has made you gun shy about giving. God still expects us to do our part and to do it with the right attitude. An investment in God's Kingdom yields great dividends.

OCTOBER 30   GANGS CAN CAUSE A BANG!

Gangs are a major problem in the many cities of America. Teens and young adults are looking for identity and many times end up in jail or dead.

The greatest mass collision of ships on record occurred off Newfoundland on May 27, 1945. A west-bound convoy of 76 Allied vessels was steaming slowly through a dense fog when one struck an iceberg, discovered eight others nearby, and gave the alarm. Instantly, the entire convoy swerved sharply with the result that 22 of the ships collided with one another in the following ten minutes. Yet none sank and no lives were lost. Incidentally, it happened on the last day that vessels crossing the Atlantic were required to sail in convoy.

Following the wrong leader can lead to a devastating conclusion. Who's leading you today? Write down ways you can follow the right path and who you can lead to the right destination.

OCTOBER 31   HOLY DISARMAMENT

The University of Alabama Crimson Tide football team squared off with the Miami Hurricanes January 1, 1993 in the Sugar Bowl. This game would decide the National Championship between No. 2 Alabama and the No. 1 team in the nation, Miami. Everyone favored Miami to win with their tough defense and passing offense led by Heisman Trophy winner Gino Terretta. However, the Tide's defense disarmed the arm of Terretta and easily defeated the Hurricanes 34–10.

Our commander in chief, Christ, has defeated our enemy and declared us to be champions through him. We read in Colossians 2:15, *"And having disarmed the powers and authorities, he made a public spectacle of them, triumphing over them by the cross."*

Even if the forces of Satan seem powerful or overwhelming, we must remember that Jesus Christ has disarmed those powers. No matter what you may face today, you are victorious through Him.

It is one of the toughest physical challenges in North America. The body's endurance is put to the ultimate test. It consists of 2100 miles of trail that starts in Georgia and ends in Maine. The Appalachian Trail attracts hikers from all over for casual strolls, picnics and breathtaking scenery, but few tackle its full course. Doing so requires a person to walk step-by-step for some six months over mountains, rivers, rocks and fields.

In a way, it wounds much like the Christian life, often referred to as our "walk with the Lord," a walk that leads us over mountains, rivers, rocky times, fields and often beautiful scenery. 2 John 6 says, *"And this is love, that we walk in obedience to his commands."* This implies a daily, consistent, step-by-step test of endurance. Wherever your walk with the Lord may lead, be sure to stay on the right path and keep putting one foot in front of the other.

Read Genesis 3:1-14.

Recently while traveling in Hollywood, California, I rode down Hollywood Boulevard. I saw Manns Chinese Theater, the stars' footprints and handprints, The Roosevelt Theater, The Directors Guild, and about every sin known to man. Homosexuality, prostitution, hatred, anger, addiction, and the list goes on!

The entertainment industry has saturated America with a New Age philosophy that says, "Do what thou wilt will be the whole of the law." Satan told Eve to take the fruit and her eyes would be opened. It's still the same today. Satan dangles the fruit among America's colleges and universities. Drunkenness, promiscuity, fornication, lying, stealing, cheating, etc. Do not take the fruit of the world; take only the fruit of the Spirit.

Have you ever met a person who was always in a good mood all the time? They either cheer you up or drive you crazy. Jill and Megan were roommates at Oklahoma State and enjoyed a great friendship with each other. Both were outgoing and social, but once inside the dorm room, Jill cherished a few quiet moments, especially in the mornings. Megan, on the other hand, woke up singing and talking, regardless of the pressures that often were felt in college. Did Jill eventually take out a contract on Megan? No, actually, she began to notice that she started each day in a better mood as a result of Megan's joy.

In Acts 16, Paul and Silas are in prison. They have been beaten severely and thrown into a cold, damp dungeon along with some other men. Verses 25-26 tell us that at midnight, in the darkest hour, Paul and Silas began singing. The other prisoners were listening, when all of a sudden an earthquake shook the prison, the doors flew open, and everyone's chains fell off! If we can learn to praise God even in dismal times, to make a joyful noise, it will impact our lives and those around us in unforgettable ways.

Read Esther 8:5-8.

Perception is quite important. Regardless of how you feel personally, it should be your goal that those around you have a good perception of you. Read the following story of a mother and her children:

"What are you children playing?" asked the mother.

"Church," chorused the crowd of youngsters.

"You know that you shouldn't whisper in Church."

"Yes, but we're the choir."

The little ones above perceived the church choir in a certain way. How do your friends (especially those who are unsaved) perceive you?

Take time to examine your Christian life. Are you living a consistent witness on your campus? Prayerfully consider your daily routine and walk. It is God's wish that others perceive you as a spirit-filled child of God.

NOVEMBER 5   IT'S NEVER TOO LATE FOR GOD

John 11 tells about anincredible miracle of Jesus that is fun to read over and over. Yet there are great truths surrounding this impossible situation that need to sink deep into our hearts. What is something in your life that you consider to be impossible? A family situation, Calculus, a drug habit? The hopelessness that comes with an impossible situation can seen overwhelming. That is how Mary and Martha, Lazarus' sisters, felt. In verse 32, Mary falls at Jesus' feet and says, *"Lord, if you had been here, my brother would not have died."* Jesus was too late, Lazarus had been dead four days.

Maybe you feel your situation is too far gone, that there is no hope or no chance of life. Anytime Jesus shows up on the scene, there is hope. He shouted life into that grave, and Lazarus was raised from the dead. Jesus can bring life and hope into your impossible situation today! Simply ask Him.

NOVEMBER 6   THE SCHOOL OF HARD KNOCKOUTS

One of the toughest things we deal with in life on a daily basis is temptation. Do you know anyone that enjoys the battle that takes place when you are tempted? Not! Imagine what pain you would live in if the only way of learning anything in life would be to get in a boxing ring with George Foreman. Every lesson had to come in the form of boxing. "Okay," your dad would say, "you are going to learn how to respect your mother! Get in the ring. Go to it, George."

Crazy, huh? Well our heavenly Father does not want us to learn every lesson in life by being thrown into the ring of temptation. In fact, while teaching the disciples to pray, Jesus said in Matthew 6:13 to pray this, *"And lead us not into temptation, but deliver us from the evil one."* God does not want you to learn everything through temptation, but rather through His Word and His voice. Pray this prayer every day, "Lord, lead me away from temptation."

NOVEMBER 7   TWO STEPS FORWARD, THREE BACK

Read Psalm 139:5-7.

Do you sometimes feel that every time you make progress you seem to fall further behind?

Inscribed across an old map of Jamaica is the title, 'Land of Look Behind.' The map goes back to the days when there were slaves. When the slaves escaped, they headed for the mountains. The government would send troops after them. So they frequently looked fearfully over their shoulders. This gave the mountainous area its unusual name.

This is a strange but true story. It has significance for believers. Hopefully, it's a reminder of what not to do. As spirit-filled Christians, we are not to look behind but press on ahead.

Today, do not allow yourself to look at past mistakes. Instead, press on, look ahead, get excited about your future with God!

NOVEMBER 8   DEPRESSED OR DELIGHTED

Read Zephaniah 3:16-17.

We are called to be the light of the world. We are to radiate God's love. Are those you come in contact with each day being changed due to your countenance, love, joy, and peace? They should be! As Hudson Taylor said: "If your father and mother, your sister and brother, if the very cat and dog in the house, are not happier for your being a Christian, it is a question whether you really are." As a Christian, Jesus desires to shine His light through your face, your words, your actions and your smile!

NOVEMBER 9   PURSUE HIS PURPOSE

Read Acts 13:36 and ask yourself, "What is my purpose on this earth?" David Livingstone is known as the most famous missionary of the Gospel in the past 500 years. Soon after arriving in Africa, a huge lion almost ripped his arm from his body, leaving him permanently crippled. He spent more than 5 years separated from his wife and children.

His health declined to the point that he had no strength, his feet were covered with boils, and his teeth had fallen out. Unable to stand or walk, he commanded his friends to carry him from village to village, where he could preach while propped up on a stretcher. David Livingstone was found one morning kneeling beside his bed where he had died during the night. He spent 39 years in Africa, traveled 29,000 miles and preached the Gospel to over 2 million Africans. Acts 13:36, speaking of David from the Old Testament, says he *"served God's purpose in his own generation..."* You have a purpose. Are you carrying out God's purpose in your life?

NOVEMBER 10   HIDE AND SEEK

Read 2 Samuel 19:24-28 and 21:7.

Do you remember playing hide and seek? The anxiety was so strong when you knew you were about to be found.

Engaged in a series of services in a church in California, a pastor looked at the bulletin and singled out the following announcement, which was all right except for the lack of one 'g'. "Choir rehearsal this afternoon at 3:30. Everyone who wishes to sin in the choir must come to practice."

A hidden or missing letter can cause much confusion. A hidden or missing servant for Christ can mean misguided sheep looking for a shepherd. Are you hiding or missing in action? Write down how you can be seen or get involved in ministry today.

NOVEMBER 11 — WHAT A DYNAMITE IDEA!

Can you imagine what it would be like to wake up one morning and read your own obituary? That happened in 1880 to Alfred Nobel. Actually, a reporter noted Alfred as the one who had passed away when in reality it was his brother. As he read about his own life, something troubled him greatly. The only thing the paper said was that he was responsible for inventing dynamite. Not wanting to be remembered for creating a killing machine, he put his money into a fund to reward people who helped bring peace to the world. We now have the Nobel Peace Prize.

In Matthew 5:9, Jesus said, *"Blessed are the peacemakers, for they will be called sons of God."* What a reward that is. You may never achieve fame for creating peace, but you will be blessed in God's eyes. What are some simple things you can do today to bring peace to those around you?

NOVEMBER 12 — A NEEDLE IN A HAYSTACK

Read Haggai 2:6-9.

Are you resting in God's perfect plan for your life? Do you realize that He has given Himself to you and that He is all you really need? Ponder this:

In his journal, John Wesley relates this incident: "Today I visited one who was ill in bed. She had buried seven of her family in six months, and had just heard that her beloved husband was cast away at sea. I asked, 'Don't you fret at any of these things?'

She answered with a loving smile on her pale cheeks, 'Oh, no! How can I fret at anything which is in the will of God? Let Him take all besides; He has given me Himself. I have learned to love and praise Him every moment.' "

Peace is often hard to find in this world, similar to a needle in a haystack, but you can rest in Jesus today!!

NOVEMBER 13   IT'S A MATTER OF TIME

Procrastination. Do thoughts of term papers and reading assignments flash through your head when you hear that word? One college student said, "Why do today what you can put off until tomorrow?" We often fall into the habit of procrastinating, but one thing is for certain: winning the lost cannot be put on the back burner. John writes in John 4:35, *"Do you not say, 'Four months and then the harvest?' I tell you, open your eyes and look at the fields! They are ripe for harvest."* We cannot allow Satan to sidetrack us or get our focus off of the lost. Now is the time to act.

A young girl visited a farm and wanted to buy a large watermelon. The melon was $3, but she only had 30 cents. The farmer pointed to a small watermelon in the field and said, "How about that one?"

"Okay, I'll take it," said the girl. "But leave it on the vine and I'll be back for it in a few months." The time for harvest is now. Just do it.

NOVEMBER 14   PINOCCHIO'S PROBLEM

Read Isaiah 53:5.

Do you remember how Pinocchio's nose would grow every time he told a lie? Can you imagine what life would be like if every time we told a lie our nose would literally grow three inches in length?

As with Pinocchio, we receive consequences to our actions of wrongdoing. There are many opportunities to sin or miss the mark. It was our sin that nailed Jesus to the cross, and maybe it was that sin that hurt Him the most! Are there any sins that you need to repent of today? Jesus is listening!

NOVEMBER 15   THE THEATER OF YOUR MIND

On October 21, 1992, an event took place that gained world-wide attention. No, it was not a new medical discovery nor was it a natural disaster. On that Wednesday, Madonna's new, 128 page, book entitled *Sex* hit the bookstores all over the world. The book featured 401 pictures in all, 75 in which Madonna is nude. The book, which is controversial to say the least, sold for $49.95. It also featured S & M, lesbian and homosexual-type poses and other sexually-explicit themes and fantasies.

The human brain is much like a sponge, absorbing images and principles constantly. The principles and ideas that are soaked up by the brain are the same principles, thoughts and ideas that will surface when the pressure is on. Philippians 4:8 says, *"...whatever is true, whatever is noble, whatever is right, whatever is pure, whatever is lovely, whatever is admirable—if anything is excellent or praiseworthy—think about such things."* What type of images are being absorbed by the theater of your mind? Put on the helmet of salvation and guard your mind with all diligence.

NOVEMBER 16   ONE SHARP SWORD

Read Proverbs 15:1-4.

As you read this passage, think about the tongue and how you use it when speaking to others.

Nathan Faris, a seventh grader in DeKalb, MO, went to class one day in 1987. He put an end to merciless teasing that he had received for years. Even as he pulled the .45 caliber automatic handgun out of his duffle bag, the other kids in his class teased him. Nathan warned them that it was a real gun, but they continued, saying, "Yeah, real plastic!" Then Nathan squeezed the trigger and a real shot rang out. By the time the ordeal was over, Nathan had killed a classmate and shot himself in the head. This tragic event was the culmination of years of pain. Insensitive kids had teased Nathan too long. What would the result of his life be if they had used kind words instead?

Words are powerful. They can be used to encourage or they can tear down. As Proverbs 15:4 says, *"A deceitful tongue crushes a person's spirit."* Find someone to encourage today.

NOVEMBER 17   JUST LIKE THE REAL THING

Are you a person who has expensive tastes and enjoys the finer things in life? Well, chances are you are like most people and cannot afford the real thing. Never fear. Imitations are everywhere. One thing the world will not tolerate, however, is a fake Christian. 1 John 3:18 says, *"...let us not love with words or tongue but with actions and in truth."* We must be genuine and real in our walk with God and our love for others.

There were two taxidermists standing in front of a store window, criticizing an owl that was on display. It was not mounted properly, its eyes were not natural, the wings were not in proportion with its head, the feathers were not neatly arranged, and its feet could certainly be improved. When they finished their critique, the owl turned its head and winked at them. You may not be perfect, but you can at least be real.

Did you know that man can live about 40 days without food, about 3 days without water, and about 8 minutes without air? Scientists and doctors have discovered this in regard to our physical bodies, but remember that our spiritual bodies also need food, each minute of every day. Psalm 121:1-2 says, *"I lift up my eyes to the hills, where does my help come from? My help comes from the Lord, the maker of heaven and earth."*

Today, look to the One who gives you daily life and help in every situation. Keep your spiritual body healthy. Do not let Satan rob you by tempting you to starve.

NOVEMBER 19   THE MORE WE WORK TOGETHER...

A county fair sponsored an event called a horse-pull. Various weights are put on a sled, hitched to a horse and pulled along the ground. The grand champion pulled a sled with 4500 pounds on it. The runner-up pulled 4200 pounds. Some men began speculating what both horses could pull if they were hitched together. Individually, the combined weight they could pull was 8700 pounds, but when they were hitched together as a team, they pulled over 12,000 pounds!

Christians can learn a valuable lesson from this story on the importance of working together. Paul said to the Church in Philippians, *"...stand firm in one spirit, contending as one man for the faith of the gospel." (1:27)* There are no lone rangers in the body of Christ. We are a team and must work together each day to impact this world.

Read 1 Samuel 26:8-9.

Do you remember fighting to get in line in elementary school? It was hard to let others go ahead.

One day, in the Mayo Clinic, an affluent and obnoxious newcomer spied a white haired doctor standing in the lobby. He strode up officiously and said, "Tell me, my good man, are you the head doctor here?"

Dr. Will, elder of the two famous Mayo brothers, bowed courteously to his interrogator. "No, kind sir, it must be my good brother you are seeking. I am the belly doctor."

In 1 Samuel, David could have killed Saul. Instead, he preferred him and respected him for his position and age. It is God's will that we honor those where honor is due. How can you honor and bless someone today?

The miracle of salvation is truly a miracle. God is somehow able to cover the ugliness and filth that sin creates in us with the blood of Jesus and turn us into something beautiful and, in His eyes, priceless. We are all born sinners. Romans 3:23-24 says, *"For all have sinned and fall short of the glory of God, and are justified freely by His grace through the redemption that came by Christ Jesus."*

The story is told of a king who possessed a very valuable diamond. By accident the stone got scratched. The king was very upset and called for the court jewelers to repair the scar. They couldn't. A very wise man stepped forward and asked if he could try to repair the stone. The king consented and in a few days, he returned with the diamond. He was unable to erase the scar, but instead had etched onto the stone an image of a rose, using the scratch as a stem.

Read Hosea 4:6-13.

Divorce is as commonplace today as eating at McDonalds. Almost 6 out of 10 marriages end in divorce . Those who lose the most is the kids. The last thing on a newly married couples mind is divorce. Everything seems blissful, happy, joyous, passionate and eternal, yet many split to go separate ways.

If ever someone had a right to divorce it was Hosea. He had a wife given to many others and could have said the 'D' word, but he didn't. He held on until she returned to love and submit to the covenant of marriage. At salvation, one makes a covenant of marriage to God. He never leaves! If we're not careful though, other lovers will prostitute our affection. Entertainment, careers, friends and habits can pull us away. Have you spiritually divorced God? Are you running to other lovers or remaining true to your first love? Marriage still works with God. No matter what Donahue or Geraldo say, never say the 'D' word!

NOVEMBER 23   I THINK I CAN, I THINK I CAN, I THINK...

Determination is a key factor in making it as a Christian in a non-Christian world. Often, at the first sign of opposition or adversity, people are quick to throw in the towel in the Christian life. We must determine that no matter what the cost, no matter how tough it may be at times, we are committed to the cause of Christ for the long haul. Philippians 4:13 says, *"I can do everything through him who gives me strength."*

Ever hear of Bob Wieland? He competed and completed the Los Angeles Marathons in 1987 and 1988. He competed and completed the New York City Marathons in 1986 and 1987. He competed and completed the Race Across America on a specially built bicycle twice. He shattered the world record in the bench press four times, and he walked 2,784 miles across America to raise funds for world hunger. Quite a list of accomplishments, huh? Especially when you consider the fact that Bob has no legs! He took 4.9 million steps walking on his arms. You can make it as a Christian!

NOVEMBER 24   I'LL JUST RUN AWAY

College life can make you want to scream! Read Jonah 1:1-3. Do you ever feel like running away? Obviously, many do. Last year, over 1 million young people ran away from home for the first time. We live in a quitting society. If we don't like our job, we just quit. If we don't like our spouse, we just quit. If we don't like our preacher, we just quit to that church. If we don't like school, we just quit.

Are you a quitter? Are you running from responsibility? Winners never quit and quitters never win. Running away, as Jonah discovered, is not the answer. There are other alternatives. Run to God. He is the Answer!

NOVEMBER 25   SWEET DREAMS

A man and his wife were leaving church one Sunday morning. "Did you see the color of the dress on the woman in front of us?" the wife asked. "Or the red dress on the woman across the aisle? And the suit the pastor had on?" "No, I didn't," replied the husband. "I'm afraid I dozed off." "Well, a lot of good church does you!" she snapped.

Read 1 Thessalonians 5:6. Many things in today's world can lull us to sleep if we are not careful. As a soldier in God's army, we have a duty to stay alert and not allow ourselves to become dull or apathetic. *"So then, let us not be like others, who are asleep,"* the scripture says, *"but let us be alert and self-controlled."*

Remember when you got in fights with a sibling or friend and a grown up said, "Tell him your sorry!" Sometimes sorry isn't enough. Read 2 Chronicles 7:14-15.

Repentance is not a word used often today's vernacular. However, it was the first message preached on the day of Pentecost. Peter stood to say, "Repent, for the kingdom of heaven is at hand!" Repentance is different from saying, "I'm sorry." Sorry means, "Oops, I got caught." Repentance means turning from and walking the opposite direction, a U-turn if you will.

What's wrong in America's universities and colleges is not the devil's fault, it's the churches fault. If we will humble, submit and pray and turn from wickedness, God will send healing and forgiveness.

Are you saying, "I'm sorry," but not repenting? Ask yourself what you need to repent of and watch God send His healing and blessing as you do it.

NOVEMBER 27   LITTLE BECOMES MUCH

When you read the miracle Jesus performed when He fed five thousand followers in John 6, you begin to understand how His hand on our lives can affect us and this world. As dusk was approaching, Jesus' disciples urged Him to send the people away so they could go home and eat. Andrew spoke up, obviously having great faith in Jesus, and said there was a boy there with five loaves of bread and two fish. In verse 11, Jesus takes the items , blesses them, and distributes them to those who were present. Five thousand people were fed, and there were even leftovers.

So often we feel that what we have to offer God is small and insignificant. Maybe you don't have a beautiful singing voice or an outgoing, charismatic personality. God is not looking for ability, but availability. Offer God whatever you have. As we see in John 6:11, little becomes much when placed in the Master's hands.

NOVEMBER 28   PRAISE HIM WITH EVERYTHING

It is difficult to believe but Givonni Battista Rubini, Italian operatic tenor, is noted for the fact that he sang one high note with such force that he broke his collarbone.

Psalm 103:1 says, *"Praise the Lord O my soul and all my inmost being, praise His holy name."* Rubini definitely sang from his soul and entire being. How it should be in our daily walk with God that we sing and praise Him from ours, too. We should continually live Psalm 103:1. The spirit-filled life is to be one filled with praise. Think of Rubini's forceful singing and enter into forceful praise.

NOVEMBER 29   THE BREAKFAST OF CHAMPIONS

Someone has said that a dusty Bible invariably leads to a dirty life. Without a daily, consistent dose of God's Word, sin is bound to creep in and overtake us. God's Word is our spiritual food. If you did not eat anything for a week, your body would begin to lose strength and would be more susceptible to disease.

Hebrews 4:12 says, *"For the Word of God is living and active. Sharper than any double-edged sword, it penetrates even to dividing soul and spirit, joints and marrow; it judges the thoughts and attitudes of the heart."* If we fail to feast on God's Word, our spiritual food, we will become weak and sin-sick. Eat your Wheaties, but don't forsake the Word of God.

NOVEMBER 30   REJOICE, REJOICE O CHRISTIAN!

Luke 10:19-20 give the Christian two of the most uplifting, encouraging facts a person could ever want. Fact one–You are a winner. Fact two–You are going to heaven. It feels so good when someone puts you in charge of an event or gives you some important task or responsibility. The authority that comes with being in charge builds a person up. It may come in the form of a job promotion, an appointment, an election to a position, or a God–given responsibility.

Jesus says in verse 19, *"I have given you authority to trample on snakes and scorpions and to overcome all the power of the enemy..."* We have authority over Satan and are winners as a result. Jesus also says in verse 20, *"...do not rejoice that the spirits submit to you, but rejoice that your names are written in heaven."* We must use the authority Christ has given us. We must also keep in perspective our greatest joy, heaven.

Remember the old television show entitled, "The Untouchables?" We now have a new version being played out everyday in America. The new version has nothing to do with detectives or mobsters, however. It involves the countless number of people in our country that society deems as "untouchable." The homeless, drug addicts, AIDS victims, rebellious teens—all resemble them as spoken of in Luke 10:30.

This man was traveling from Jerusalem to Jericho when he was suddenly mugged and beaten by robbers. A preacher saw the man but considered him to be untouchable. He crossed over to the other side of the street as he passed by. Another man, a Levite, did the same, but a Samaritan man reached out and touched the untouchable. He bandaged his wounds, cleaned him up, fed him and put him up in a nice hotel. The Samaritan crossed religious and racial boundaries by showing love to this Jewish "untouchable." Does Jesus expect the same from us? What do you think? Let's reach out and touch someone.

DECEMBER 2 **WORSHIP IS A WEAPON**

Read Psalm 135:1-7.

During the 1980's, the Pentecostal Charismatic Movement put praise and worship at the forefront of evangelical ministry. Now mainline denominations are involving their congregations in praise and worship. Why? Because worship is a weapon.

Napalm is a man-created heat that reaches temperatures up to 2060 degrees C. Its burns are deep and extensive and result in severe scars.

Napalm can burn the flesh and so can worship. Worship puts you in touch with the spirit realm where you can meet God one to one. Take a moment to crucify (or burn) your flesh and attack heaven with your weapon of worship.

One of the great benefits of the body of Christ is that a person never has to face trials or difficulties alone. They can choose to do that, of course, but there are always people close by to help lighten the load. This is exactly what Paul is referring to in Galatians 6:2 when he says, *"Carry each other's burdens, and in this way you will fulfill the law of Christ."* We, as Christians, have a responsibility to help lighten one another's load. The burdens come in all shapes and sizes. It may be stress from a heavy class schedule, pain from losing a mom to cancer, or heartache from being dumped by a guy or girl. Whatever, the case, we must find ways to help one another through difficult periods in our life.

Take, for example, a group of Jr. High boys in Illinois who found out their classmate had been diagnosed with leukemia. They knew their friend Mark would be embarrassed over losing his hair during chemotherapy treatments. So in order to bear his burden, some 15 seventh- and eighth- grade boys shaved their heads bald. Mark's heavy load became lighter.

DECEMBER 4   YOU GOTTA BE FAT!

Read Proverbs 14:1-6.

The infatuation that "thin is i," makes diet and exercise programs flourish. However, the scripture tells us to be fat, that is, faithful, available and teachable. Society says that beauty is on the outside. God says that beauty is in the heart.

When the citizens of High Wycombie, England, elect a new mayor, all the town councilors are weighed in public, following an ancient custom. Those whose weight is less than or at least no more than when they took office are warmly applauded; they have not grown fat at public expense.

I am not advocating overweight as God's best value. I am saying that God is not as concerned with the outward appearance as He is with the heart. Knowledge minus obedience is deception. God wants us faithful, available, and teachable on the inside.

P.S. Watch those calories too. It's good sense!

DECEMBER 5   BEATING THE BLUES

Life is not fair. Life often delivers unexpected blows that are difficult to handle. It drops things at our doorstep that go beyond our ability to cope. Why the brain tumor? Why the accident? Why the closed door of opportunity? Questions arise that many times we will not be able to answer, nor will anyone else. God is still God in the midst of unfairness and difficulty! In 2 Corinthians 12:9 it says, *"...'My grace is sufficient for you, for my power is made perfect in weakness. Therefore, I will boast all the more gladly about my weaknesses, so that Christ's power may rest on me.'"*

On February 25, 1989, after 29 seasons as head coach of the Dallas Cowboys, Tom Landry was suddenly and unexpectedly fired with no warning. In his autobiography, speaking about his faith in God, Landry says, "It's that belief, that faith, more than anything else, that enabled me to last twenty-nine years on the sidelines of the Dallas Cowboys. It's that faith that has allowed me to keep my perspective and not feel devastated or bitter about being fired. And it's that faith that gives me hope for whatever the future holds..."

DECEMBER 6   OPEN YOUR MOUTH

Read Exodus 4:15-17.
Do you remember as a child taking medicine? Did you want to run, cry, and throw a fit? A mouthful could make you sick.

The Governor of Iowa has a name that makes rather a neat mouthful, Bourke Blakemore Hickenlooper. He himself tells about a drugstore clerk who refused to charge ten cents' worth of asafetida to the Hickenlooper Account. "Take it for nothing, boss," the clerk said. "I wouldn't write both asafetida and Hickenlooper for a dime."

Sometimes it is hard to open our mouth and proclaim Jesus as Lord in a public place. The scripture says to open your mouth so that God can fill it. Knowledge minus obedience is deception. If God said shout it from the rooftop and we do not do it, are we obedient or deceived? Can you open your mouth for God to fill it today?

DECEMBER 7   PURSUE HIS PRESENCE

According to Webster's Dictionary, to pursue means to follow in an attempt to capture. For example, a defensive back in football pursues the man with the ball in an attempt to capture or tackle him. Everyone in life is in pursuit of something. In Luke 8:42-44 we read of a woman who was in pursuit of the presence of Jesus.

It says, *"...As Jesus was on his way, the crowds almost crushed him. And a woman was there who had been subject to bleeding for twelve years, but no one could heal her. She came up behind him and touched the edge of his cloak, and immediately her bleeding stopped."* This woman we read about was determined to get close to Jesus and to even touch Him, despite the enormous crowd that surrounded Him. Her efforts were rewarded. The moment she touched Jesus, she was instantly healed. If we pursue Christ with the same determination to be in His presence and to touch Him, we will be rewarded in phenomenal ways. Let us capture His presence in our lives!

DECEMBER 8   WHAT'S YOUR PLAN?

It has been said that if you fail to plan, you plan to fail. Are you planning for your future? We have daytimers, elaborate calendar systems, cellular phones, and modern fax machines for the "on the go" society of the 90's. Read Micah 4:6-15.

The Lord has a plan for you! You can walk in victory and not defeat, blessing and not cursing, to be the head and not the tail. God has a beautiful plan for your life. There is no demon, no government, no institution or administration that can cause you to miss God's plan for your life. The only person that can cause you to miss God's plan is you by being more preoccupied with life as you plan more than what God has planned for your life.

Drugs are getting worse; suicide, runaways, teen pregnancy, AIDs, politics, global economics are all getting worse. Do not fear, God has a plan for you to have peace and joy and patience, goodness in the middle of all of life's struggles. He's the master architect. Let Him plan your life.

DECEMBER 9   WE DIDN'T START THE FIRE

The Billy Joel song of the 90's was making a statement, "Don't blame us for the world's problems. We didn't start the fire." In other words, it's not my responsibility. Yet the Bible makes it clear that we do have a responsibility to stand for what is right and to do what we can to correct what is wrong. Jesus said in Matthew 5:14, *"You are the light of the world."* Then in verse 16, He says, *"...let your light shine before men..."*

A small town fire department was holding a pancake breakfast to raise money for equipment. One of the volunteers asked a local businessman to buy a ticket. "I don't eat pancakes!" the man said angrily.

"And we don't start fires," the fireman shot back. Maybe you are not to blame for the world's fires or problems, but God is looking to us to douse the fires with the water of His Word and to bring hope and restoration.

Read Lamentations 3:22-25.

Do you ever stop to consider the grace of God? A young boy once wanted a new bicycle. His father had told him that their family could not afford it. The boy's neighbor had the exact one that he wanted. One night, the boy slipped out of the window and went to steal the bicycle. While taking the bike, the neighbor's dog began to bark and awakened the family. They caught the boy in the act! Although the boy pled for mercy, the juvenile court was called, and the boy was sentenced to 13 weeks in juvenile detention. Judgement came instead of mercy.

Jesus is the great giver of mercy. Since He died on the cross, we receive, as a free gift, mercy from the greatest Judge of all! There is no sin, no mistake or failure that God cannot forgive. Do you need mercy? Call out to God now and you will find that His mercy endures forever!

DECEMBER 11   ARE YOU MAKING A FASHION STATEMENT?

Without realizing it, Andrew Martinez gave Christians in America a graphic example of how foolish it is to not be ready for battle. Martinez, a student at the University of California at Berkeley, made it a practice in the Fall of 1992 to walk around campus totally nude. He jogged, ate in the dining halls and attended classes while naked. It was his way, he said, of protesting sexually repressive traditions in Western society.

Crazy isn't it?! Unfortunately, many Christians walk around every day as "spiritual streakers," naked when it comes to being prepared or dressed for battle. We must, as Ephesians 6:13 says, *"...put on the full armor of God, so that when the day of evil comes, you may be able to stand your ground..."* Putting our clothes on in the morning is a priority (unless you are Andrew Martinez). Putting on God's armor in the morning is vital for survival. It is a must for the Christian.

DECEMBER 12   A DAY OF INDEPENDENCE

Read Nehemiah 6:15-7:2.

The 4th of July is a significant day in American history. It is the day we celebrate independence from all other nations to be a sovereign nation to ourselves.

Nehemiah had built the walls of Jerusalem for they were in ruins (Nehemiah 2:17). After 52 days, the walls were completed and a celebration followed because their enemies left in fear. I'm sure they had fireworks of sorts.

On the day you asked Christ into your heart, you declared independence from the devil and sin. On the day that you receive the baptism of the Holy Spirit (Acts 2:4), you declare independence of fear to be a bold, zealous witness for Christ. Are you independent or dependent? Write your declaration of independence today!

DECEMBER 13   WHICH PASTURE ARE YOU PARTYING IN?

If you want a lesson in human behavior, watch a herd of sheep. Where one goes, they all go. Maybe that is why Jesus referred to us as sheep. Mark Twain said, "We are discreet sheep; we wait to see how the drove is going, and then go with the drove." Better yet, watch students at most any university or college campus. Students are determined to be different and not conform to anyone else's standards or opinions, while the whole time they are following the drove without realizing it.

The Bible addresses the issue of conformity in Romans 12:2. It reads, *"Do not conform any longer to t he pattern of this world, but be transformed by the renewing of your mind."* How do we renew our minds? By washing our minds daily with the water of God's Word.

As we adopt the mind of Christ, we will become more like Him and less like the world, conforming to His image of righteousness, purity, and love.

DECEMBER 14   HURRY, HURRY, HURRY

Composer Igor Stravinsky's publisher urged him to hurry the completion of a new composition.

"Hurry!" he cried angrily. "I never hurry. I have no time to hurry."

Funny, but true. Many times college students find themselves in Stravinsky's position in life. Hurry, quick, complete this paper, project and study guide. Study for this test. Attend this social function. Take time for a daily devotion, and the list goes on!

Even though you find yourself at a busy stress filled time in your life, slow down and reflect on Jesus and read Proverbs 4:20-27. You will make it . You not only want to be a finisher, but also want to see that you remained focused in your Christian walk. These verses in Proverbs are some you should keep handy and read often. Possibly, the Lord is speaking to you to memorize them.

DECEMBER 15   LET GO AND LET GOD

A man was peering over the edge of a cliff when he suddenly lost his footing and fell over. As he was falling, his hand grasped a small branch sticking out of the side of the cliff, stopping him but dangling him over the rocks below. He cried out, "Help! Is anybody up there?"

A voice replied, "I am here."

"Who's there?" he said.

"It is I, the Lord," the voice said. "Have faith. Trust me and let go."

After a long pause, the man shouted back, "Is anybody else up there?"

How many times have you been in a position where God asked you to do something, but you did not want to obey? Maybe your faith went out the window when God asked you to witness to your roommate, or to invite a friend to your church's weekend retreat. God will never ask us to do anything simply to make a fool out of us or to let us fall flat on our faces. Hebrews 11:6 says, *"And without faith it is impossible to please God, because anyone who comes to him must believe that he exists and that he rewards those who earnestly seek him."*

DECEMBER 16   FALSE ALARMS?

A Wisconsin town council was discussing how to dispose of an old fire truck, now that they bought a new one. One counsellor finally stood: "Use the old, inadequate truck to answer false alarms."

Let's be careful so that we don't have this type of mentality. Could the counsellor know "false alarms" ahead of time? Absolutely not!

In your spirit-filled Christian walk, you need to know what thought to dispose of and what is from your heavenly Father. Perhaps in some of your classes, you are hearing new ideas, thoughts or philosophies or old ones presented differently. Pray for discernment each day. You need not be like the Wisconsin town council. There are no false alarms with Jesus.

Let's strive to be like Moses and Aaron in Exodus 7:6-7. Just obey God, no matter what the time in your life may be. Obey Him and your questions will surely be answered.

DECEMBER 17   TAKE TIME TO MAKE TIME

"I just need more hours in the day," Mike said to his friends at the Bible study. "Between studying and working part-time, I never find time to read God's Word, pray or get involved with a church."

Have you ever said those same words or some similar? The clock is often our worst enemy. Jesus said in John 9:4, *"As long as it is day, we must do the work of him who sent me. Night is coming, when no one can work."* Somehow, our walk with God and our service to Him must take top priority in our lives.

Bill, a Chicago businessman, certainly does not have extra time or hours in the day, yet one day in 1984, after visiting a Chicago public school, Bill determined that he would make time to make a difference. Bell Kellogg, great-grandson of the founder of Kellogg cereals and president of General Packaging Products Inc., set aside one day out of every week to teach English at the school. God has given each of us opportunities to make a difference. While we can, let's make time to pray, read, and work for God.

DECEMBER 18   DON'T BE A FOOL

Michigan State's football coach Duffy Daugherty received a letter addressed to "Duffy the Dope." "Didn't that make you mad?" he was asked. "I didn't mind getting the card," he said. "It was pretty funny. The thing that bothered me was that the East Lansing post office knew exactly where to deliver it."

Ecclesiastes 7:9 states that we are not to be quickly provoked in our spirit because anger resides in the lap of fools. Do you think "Duffy the Dope" knew Ecclesiastes 7:9?

During your college experience, you will have opportunities for anger to rise within you. Professors, dating relationships, campus life, etc. Know that aggravating circumstances will come your way, but heed Ecclesiastes 7:9.

Pray this simple prayer: "Dear Lord, I thank you that my college experience will be blessed and ordered by You. Help me not to be quickly provoked. I do not want to keep anger inside nor do I want to be foolish. Thank you, Father, for helping me to live Your Word."

Ephesians makes it clear to us that we are saved by grace, not by works. James tells us that faith without works is dead. As Christians, we are charged by God to work to build His Kingdom because we are burdened for souls. Robert Frost once said, "The world is filled with willing people; some willing to work, the rest willing to let them." Unfortunately, this is also true of the church.

One lady told the story of her 4-year-old son Jonathan eating dinner at their youth minister's home. The pastor asked if Jonathan like Sunday School. "NO!" he said, "All we do is sit on our butts and color." In 1 Corinthians 3:13, we read, *"...his work will be shown for what it is, because the day will bring it to light. It will be revealed with fire, and the fire will test the quality of each man's work."* May we never be guilty of "sitting on our butts" and doing nothing for God. God will one day examine all we do for Him. Let your prayer today be this, "God, help me today to do at least one thing that will make an eternal difference for your Kingdom."

You may know that Richard Wagner's music first brought groans and laughs from the majority of critics. Now it is recognized that his compositions have changed the music world.

Your personal music may be like Wagner's. Possibly, you feel awkward singing a new song to Jesus. Maybe you feel your melodies are unimportant to Him. Take time to read and meditate on Psalm 150. This scripture ends with *"Let everything that has breath, praise the Lord."* Know that your music and praise is precious to your Heavenly Father. Just as Richard Wagner's music changed the music world, your music can strengthen and deepen your relationship with Almighty God.

"Consistency is the key," the swimming coach screamed to Robin. "Don't change your stroke or your kick. Stay consistent." Robin's times had gotten worse, not better, in spite of rigorous training and practice. Consistency is paramount to success, no matter what you are doing. One of the greatest compliments a person can receive is that he or she is consistent, especially in the Christian life.

Jesus points this out in Matthew 16:18 when He calls Peter "petros" or "rock." He says, *"And I tell you that you are Peter, and on this rock I will build my church and the gates of Hades will not overcome it."* What a compliment! We should strive to be so stable, consistent and solid in our commitment to Christ that God will build His Kingdom with us. What are some things that cause you to be inconsistent in your relationship with the Lord? Let's improve those areas of out life and become rock solid for God.

DECEMBER 22   *"I DID IT MY WAY"*

Read Genesis 4:1-7.

A group of Salt Lake City executives, after lunch, hailed a cab to go back to the office. As they climbed in, one exclaimed, "Well, back to the salt mine!" The cab bowled along and the executives became so engrossed in conversation that they didn't look up until it stopped–at the entrance to the salt mine at the edge of town.

I guess you have noticed that you get what you ask for in life. Be careful what you ask from Jesus. You just might get it! Try praying that God's will be done in your life instead of your own. He knows what is best for us at all times.

Often we persist and weep to the Lord and He lets us have "our way." It is not always His perfect will and disaster and confusion follows. Prayerfully consider His will for you!

If you examine life in the 90's, it is obvious that much occurs in any given day. For example, in an average day in America, 35 people turn 100; the Smithsonian adds 2500 items to its collections; dogs bite 11,000 citizens; the U.S. Government issues 50 more pages of regulations and we eat 75 acres of pizza, 53 million hot dogs, 167 million eggs, 3 million gallons of ice cream, and 3000 tons of candy.

Sometimes, however, the most important things in life are neglected. Hebrew 3:13 tells us, *"But encourage one another daily, as long as it is called today, so that none of you may be hardened by sin's deceitfulness."* Life boils down to relationships and we should make an effort to encourage those around us daily in their walk with God. Sure our daily schedules are littered with dogs, pizza and paperwork, yet the only thing that will remain is what we have done on this earth in our relationship with people. Think about it, today!

DECEMBER 24   IF ALL ELSE FAILS, FOLLOW INSTRUCTIONS

Read Psalm 119:105-112.

Frank Plewa was killed in a train wreck by a misunderstanding of instructions. The order read, "Hold all westward extra trains until 13:10." Conductor Duby said he read the order over the dispatcher's shoulder up to the word "westward" and, assuming the next word to be "train," rushed out to notify the "crew." That one omitted word "extra" cost the life of Frank Plewa.

Spend quiet time daily with your heavenly Father. Doing this will help you gain wisdom and complete instructions for your life.

Someone once said that there are two things in life that are certain, death and taxes. You could also include a third item in that list, rules. Many people feel that life is nothing but one big list of rules. When we begin to realize that most rules are meant to protect and not to restrict, we can begin to see that the rules God has for us in His Word were established because of His great love for us. In Hebrews 12:10 we read, *"...but God disciplines us for our good, that we may share in His holiness."*

Tim Tully, a father from New York, told the story in USA Today about the conversation he had with his 8-year-old son, Sky. They were watching one of Sky's favorite television shows. "During a commercial, " Tully says, "I gave Sky a hug, told him I loved him, and asked if he thought I was a good dad, like the TV one. He calmly said yes, so I asked why. Nonchalant, Sky replied, 'Because you set rules.'" Be thankful today for God's love and discipline.

Read Malachi 2:2.

It pays to be an effective listener. William Rathie watched with disappointment as firemen extinguished a three-alarm blaze at an abandoned waterfront pier.

"I kept telling them to let it burn," Rathie said, noting that the pier had been scheduled for demolition.

Have you ever wanted to stop listening? Lecture after lecture, professors and older classmen that think they have all the answers get difficult to listen to, but at this point in your life...be careful to listen! You may miss something of great importance.

DECEMBER 27   WHAT GOES AROUND COMES AROUND

Read Daniel 4:13-25.

In New York City, Ladder Company 25 once was called out to answer a fire alarm. While they were gone, their firehouse caught fire. Neighbors saw smoke pouring out the rear windows and phoned an alarm. Firemen from three nearby firehouses put the blaze out before Ladder Company 25 returned.

How often are we like Ladder Company 25. We leave to help with another person's problem and have a major problem break out of our own.

Reality for you could be a difficult situation ahead. Take the time, daily, to be prepared so that you are capable of overcoming anything. Notice in the illustration above that Ladder Company 25 had many to help them. The Lord will do the same for you as you give of yourself to others. Pray and believe God for Godly friends. When storms of life come, you will be glad you have them.

DECEMBER 28   "SHAKE, RATTLE AND ROLL"

Read Job 41:29-34.

During an earthquake, a few years ago, the inhabitants of a small village were very much alarmed. One old woman, whom they all knew, was surprisingly calm and joyous. At length, one of them said to her, "Mother, are you not afraid?" "No, I rejoice to know that I have a God who can shake the world."

Isn't that the way you should look at your frightening circumstances? No matter the pressure of your school load, work hours or relationships—Rejoice! God can shake the world, your awful circumstances or anything that need be. Rejoice literally means to spin around violently like a top! Right now, spin around and rejoice in God for His provision for your every need.

Read Exodus 20:3.

Terminology can be confusing as in the following story.

Admiral Dewey was a lover of children, and when he took his daily walk, always spoke to those he met.

"Well, my little man," he said to a small boy, "what are you going to be when you get to be a man?" "Oh an Animal in the Navy, just like you," replied the child promptly.

Although some terminology is difficult to understand, God's Word is not confused by terminology. Exodus 20:3 say, *"You shall have no other gods before me."* Is this easy to understand? Only Jehovah God is worth your total commitment and loyalty. Not the leisure god, the sports god, the TV god or any god. Only Jehovah God.

December 30   I'm Stuck in the Mud!

Read 2 Kings 2:1-8.

Do you ever feel that you have the river of Life inside, but your feet are stuck in the mud? Elisha felt this way when Elijah was taken up in a cloud. The followers of Elijah wanted to find him again, but Elijah's mantle had fallen on Elisha. Elisha struck the water and it parted. The people knew then that God's power was with Elisha just as it had been on Elijah. The river began to flow again, washing the mud away.

Are you in the flow of God's power and anointing? If you feel that you're stuck in the mud, ask God to wash it away so that you can flow with and for Him.

As you go through life, you will come across difficult circumstances. In the past, you may have turned to your parents, grandparents, friend or mentor for help. Now that you are becoming an adult, it's time to take a stand for yourself (if you haven't done so already), but how?

Believing in God's truths gives you the ability to stand for yourself. Acting on God's Word allows you to have the strength to stand again when the next trial comes. You will always need the support of others, but the initial stand should be taken by you along with the Word of God.

Try it. Watch God work for you and feel the joy and strength He will give you as you're obedient to Him.